PUBLISHER COMMENTARY

We print NASA's handbooks and standards for the convenience of those that use them on a daily basis. We print all of these a full 8 ½ by 11 with large text so they are easy to read. Yes, color books are expensive to print so unless the information relies on the use of color for proper interpretation or understanding, we print most books in black and white to keep the cost down. All these documents are available for download for free from NASA, however printing them all over a network printer would take days.

Why buy a book you can download free? We print this so you don't have to.

All these books are available for free download from the government web site. Some are available only in electronic media. Some online docs are missing pages or barely legible.

We at 4th Watch Publishing are former government employees, so we know how government employees actually use the standards. When a new standard is released, an engineer prints it out, punches holes and puts it in a 3-ring binder. While this is not a big deal for a 5 or 10-page document, many NIST documents are over 100 pages and printing a large document is a time-consuming effort. So, an engineer that's paid $75 an hour is spending hours simply printing out the tools needed to do the job. That's time that could be better spent doing engineering. We publish these documents so engineers can focus on what they were hired to do – engineering. It's much more cost-effective to just order the latest version from Amazon.com

If there is a standard you would like published, let us know. Our web site is www.usgovpub.com

www.usgovpub.com

Copyright © 2019 4th Watch Publishing Co. All Rights Reserved

List of Other NASA Publications Available on Amazon.com:

NASA-STD-5001B	Structural Design and Test Factors of Safety for Spaceflight Hardware
NASA-STD-5006A	General Welding Requirements for Aerospace Materials
NASA-STD-5008B	Protective Coating of Carbon Steel, Stainless Steel, and Aluminum on Launch Structures, Facilities, and Ground Support Equipment
NASA-STD-5009A	Nondestructive Evaluation Requirements for Fracture-Critical Metallic Components
NASA-STD-5012B	Strength and Life Assessment Requirements for Liquid-Fueled Space Propulsion System Engines
NASA-STD-5019A	Fracture Control Requirements for Spaceflight Hardware
NASA-STD-5005D	Standard for The Design and Fabrication of Ground Support Equipment
NASA-HDBK-8739.21	Workmanship Manual for Electrostatic Discharge Control
NASA-HDBK 8739.23A	NASA Complex Electronics Handbook for Assurance Professionals (Color)
NASA-HDBK-8719.14	Handbook for Limiting Orbital Debris (Color)
NASA-HDBK-8709.22	Safety and Mission Assurance Acronyms, Abbreviations, and Definitions
NASA-HDBK-7009	NASA Handbook for Models and Simulations: An Implementation Guide For NASA-STD-7009 (Color)
NASA-HDBK-8739.19-2	Measuring and Test Equipment Specifications NASA Measurement Quality Assurance Handbook – Annex 2
NASA-HDBK-8739.19-3	Measurement Uncertainty Analysis Principles and Methods NASA Measurement Quality Assurance Handbook – Annex 3
NASA-HDBK-8739.19-4	Estimation and Evaluation of Measurement Decision Risk NASA Measurement Quality Assurance Handbook – Annex 4
NASA RCM	Reliability-Centered Maintenance Guide for Facilities and Collateral Equipment

www.usgovpub.com

Copyright © 2019 4th Watch Publishing Co. All Rights Reserved

	METRIC
NASA TECHNICAL HANDBOOK National Aeronautics and Space Administration	NASA-HDBK-4006A
	Approved: 2018-11-28 Superseding NASA-HDBK-4006 (Baseline)

LOW EARTH ORBIT SPACECRAFT CHARGING DESIGN HANDBOOK

APPROVED FOR PUBLIC RELEASE—DISTRIBUTION IS UNLIMITED

NASA-HDBK-4006A

DOCUMENT HISTORY LOG

Status	Document Revision	Change Number	Approval Date	Description
Baseline			2007-06-03	Baseline Release Interim Standard NASA-STD-(I)-4005 was transitioned to this handbook and the standard, NASA- STD-4005. Made editorial and formatting changes. Replaced figure 10 with new figure 10 which includes "ISS Capacitance" in the legends definition area.
Revision	A		2018-11-28	Significant changes were made to this NASA Technical Handbook. It is recommended that it be reviewed in its entirety before implementation. Handbook format changed according to NASA Standards rules of formatting. The baseline version was written as one small main section with multiple appendices. Revision A does not use any appendices. Definitions were updated and clarified. Some definitions were deleted since they were no longer part of the document. The order of most sections has been changed. Notably, revision A begins with a description of the Low Earth Environment rather than describing plasmas (in general). New figure added of Daytime Ionosphere and Atmosphere Plasma Interactions (section 5); sections reordered; equations added to emphasize floating potential conditions; thermal current equations and typical values added; tether discussions enhanced with more references; current

APPROVED FOR PUBLIC RELEASE—DISTRIBUTION IS UNLIMITED

NASA-HDBK-4006A

DOCUMENT HISTORY LOG (Continued)

Status	Document Revision	Change Number	Approval Date	Description
Revision	A		2018-11-28	Continued: balance conditions explained; added example from Tribble (1995) to show how a solar array floating potential could reach 90% of the array's voltage output; removed thought experiment about two metal spheres, since Tribble reference covered that concept. Updated all International Space Station charging discussions to include information about charging measurements made by the FPMU after the initial revision of the handbook was published. Added ISO document reference for solar array arc testing and changed solar array arc discussions to conform to the standard. Added updated reference to Snapover discussion and pointed out commonalities to existing theories of snapover. Updated Parasitic Power Drain discussion. Revised Wake Effects section with additional references. Revised Arc Initiation Mechanism section and included updated references from Vayner, et al. Arc Mitigation section updated with information about a high-voltage CubeSat mission. Changed Modeling section figure from an EWB model (no longer generally used) to a currently supported NASCAP-2k model.

APPROVED FOR PUBLIC RELEASE—DISTRIBUTION IS UNLIMITED

NASA-HDBK-4006A

DOCUMENT HISTORY LOG (Continued)

Status	Document Revision	Change Number	Approval Date	Description
Revision	A		2018-11-28	Continued: Verified the public availability of all references in Appendix A. Eliminated or replaced references that were no longer available to the general public. Added references based on their introduction in the new document revision.

APPROVED FOR PUBLIC RELEASE—DISTRIBUTION IS UNLIMITED

NASA-HDBK-4006A

FOREWORD

This NASA Technical Handbook is published by the National Aeronautics and Space Administration (NASA) as a guidance document that provides engineering information; lessons learned; possible options to address technical issues; classification of similar items, materials, or processes; interpretative direction and techniques; and any other type of guidance information that may help the Government or its contractors in the design, construction, selection, management, support, or operation of systems, products, processes, or services.

This NASA Technical Handbook is approved for use by NASA Headquarters and NASA Centers and Facilities. It may also apply to the Jet Propulsion Laboratory (a Federally Funded Research and Development Center (FFRDC)), other contractors, recipients of grants and cooperative agreements, and parties to other agreements only to the extent specified or referenced in applicable contracts, grants, or agreements.

This NASA Technical Handbook establishes design guidance for high-voltage space power systems (>55 volts) that must operate in the plasma environment associated with low Earth orbit (LEO).

Requests for information should be submitted via "Feedback" at https://standards.nasa.gov. Requests for changes to this NASA Technical Handbook should be submitted via MSFC Form 4657, Change Request for a NASA Engineering Standard.

    Original signed by   11/28/2018
Ralph R. Roe, Jr.   Approval Date
NASA Chief Engineer

APPROVED FOR PUBLIC RELEASE—DISTRIBUTION IS UNLIMITED

NASA-HDBK-4006A

TABLE OF CONTENTS

SECTION	PAGE
DOCUMENT HISTORY LOG	2
FOREWORD	5
TABLE OF CONTENTS	6
LIST OF APPENDICES	7
LIST OF FIGURES	8
LIST OF TABLES	8
1. SCOPE	**9**
1.1 Purpose	9
1.2 Applicability	9
2. APPLICABLE DOCUMENTS	**10**
2.1 General	10
2.2 Government Documents	10
2.3 Non-Government Documents	10
2.4 Order of Precedence	10
3. ACRONYMS, ABBREVIATIONS, SYMBOLS, AND DEFINITIONS	**11**
3.1 Acronyms, Abbreviations, and Symbols	11
3.2 Definitions	13
4. ENVIRONMENTS	**18**
4.1 The Ambient Environment	18
4.1.1 The Neutral Atmosphere	18
4.1.2 The Plasma Environment	20
4.2 The Spacecraft-Induced Environment	21
4.2.1 Neutral Gas	21
4.2.2 Ionized Gas/Plasma	22
4.2.3 Condensable Gas	23
4.2.4 Particulates	23
4.2.5 Radiation	23
5. PLASMA INTERACTIONS	**24**
5.1 Poisson Equation	24
5.1.1 Plasma Sheath and Debye Length	25
5.2 Exposed Biased Conductors	25
5.2.1 Current Collection	26

APPROVED FOR PUBLIC RELEASE—DISTRIBUTION IS UNLIMITED

TABLE OF CONTENTS (Continued)

SECTION		PAGE
6.	**ARCING**	**37**
6.1	Solar Array Arcing	37
6.1.1	Background	37
6.1.2	Arc Initiation Mechanism – Observations and Theories	40
6.1.3	Arc Inception Voltage (Arcing Threshold)	43
6.1.4	Typical Waveform	44
6.1.5	System Response	45
6.1.6	Damage Due to Arcs	46
6.2	Structure Arcing	48
6.2.1	Sustained Arcs	51
6.3	Tether System Arcs	52
6.4	Arc-Generated Electromagnetic Interference (EMI)	53
6.5	Risks of Arcing During EVA	54
7.	**MITIGATION TECHNIQUES**	**55**
7.1	Current Collection	55
7.2	Controlling Spacecraft Potential	55
7.2.1	Positive Ground	56
7.2.2	Plasma Contactors	56
7.2.3	Encapsulation	57
7.2.4	Arcing	58
8.	**MODELING**	**60**
8.1	Spacecraft Charging	60
8.1.1	Arcing	63
9.	**TESTING**	**63**

LIST OF APPENDICES

APPENDIX		PAGE
A	References	65

APPROVED FOR PUBLIC RELEASE—DISTRIBUTION IS UNLIMITED

NASA-HDBK-4006A

LIST OF FIGURES

FIGURE		PAGE
1	Composite International Year of the Quiet Sun (IQSY) Daytime Ionosphere and Atmosphere Based on Mass Spectrometer Measurements	19
2	Voltage Breakdown of Pure Gases as a Function of Pressure Times Spacing	22
3	Electron Current vs. Bias for Three Solar Array Blanket Materials	34
4	Peak Arc Current vs. Capacitance	39
5	Arc Rate vs. Voltage for Standard Interconnect Cells	44
6	Typical Waveform for an Array Arc	45
7	Sample of Flight Array from ESA EURECA Mission after Sustained Arcing	47
8	Video Frame from EOS-AM1 Sustained Arc Test	48
9	Arc Site of Sustained Arc on EOS-AM1 Sample Array	48
10	Anodized Aluminum Plate after Repeated Dielectric Breakdown Arcing	50
11	The End of the Remaining TSS-1R Tether	53
12	EMI from a Small Solar Array Arc and a Hypothetical ISS Anodized Aluminum Arc Compared to Orbiter's Specs	54
13	NASCAP-2K Calculation of Plasma Potential and Density behind a Spacecraft at Orbital Speed	62

LIST OF TABLES

TABLE		PAGE
1	Nominal Properties of Ionospheric Layers	21
2	Leakage Current from Positively Charged Solar Arrays	36

APPROVED FOR PUBLIC RELEASE—DISTRIBUTION IS UNLIMITED

NASA-HDBK-4006A

LOW EARTH ORBIT SPACECRAFT CHARGING DESIGN HANDBOOK

1. SCOPE

The information in this NASA Technical Handbook provides an overview of present-day (2017) understanding of the various plasma interactions that can result when a high-voltage system is operated in the Earth's ionosphere, references common design practices that have exacerbated plasma interactions in the past, and recommends standard practices to eliminate or mitigate such reactions.

1.1 Purpose

The purpose of this NASA Technical Handbook is to provide design guidance referenced in NASA-STD-4005, Low Earth Orbit Spacecraft Charging Design Standard, for spacecraft electrical power systems using voltages greater than 55 volts that operate in the low Earth orbit (LEO) plasma environment encountered in altitudes up to 2000 km and latitudes between -50 and +50 degrees. Such power systems, particularly solar arrays, are the proximate cause of spacecraft charging in LEO; and these systems can interact with this environment in a number of ways that are potentially destructive to themselves as well as to the platform or vehicle that has deployed them.

1.2 Applicability

This NASA Technical Handbook is applicable to high-voltage space power systems as described in the previous section and that have been or are being designed using NASA-STD-4005. For systems intended to operate in the geosynchronous (GEO) environment, or that will operate in both the LEO and GEO environments, NASA-HDBK-4002, Mitigating In-Space Charging Effects – A Guideline, should be considered as complementary to this document.

This NASA Technical Handbook is approved for use by NASA Headquarters and NASA Centers and Facilities. It may also apply to the Jet Propulsion Laboratory (a Federally Funded Research and Development Center (FFRDC)), other contractors, recipients of grants and cooperative agreements, and parties to other agreements only to the extent specified or referenced in their applicable contracts, grants, or agreements.

This NASA Technical Handbook, or portions thereof, may be referenced in contract, program, and other Agency documents for guidance.

APPROVED FOR PUBLIC RELEASE—DISTRIBUTION IS UNLIMITED

NASA-HDBK-4006A

2. APPLICABLE DOCUMENTS

2.1 General

The documents listed in this section are applicable to the guidance in this NASA Technical Handbook.

2.1.1 The latest issuances of cited documents apply unless specific versions are designated.

2.1.2 Non-use of a specifically designated version is approved by the delegated Technical Authority.

Applicable documents may be accessed at https://standards.nasa.gov or obtained directly from the Standards Developing Body or other document distributors. When not available from these sources, information for obtaining the document is provided.

References are provided in Appendix A.

2.2 Government Documents

Department of Defense

AFWAL-TR- 88-4143, Volume 2	Design Guide: Designing and Building High Voltage Power Supplies, Materials Laboratory

NASA

NASA-STD-4005	Low Earth Orbit Spacecraft Charging Design Standard
NASA-HDBK-4002	Mitigating In-Space Charging Effects – A Guideline

2.3 Non-Government Documents

None.

2.4 Order of Precedence

The guidance established in this NASA Technical Handbook does not supersede or waive existing guidance found in other Agency documentation.

APPROVED FOR PUBLIC RELEASE—DISTRIBUTION IS UNLIMITED

NASA-HDBK-4006A

3. ACRONYMS, ABBREVIATIONS, SYMBOLS, AND DEFINITIONS

3.1 Acronyms, Abbreviations, and Symbols

°	degree
≥	equal to or greater than
>	greater than
<	less than
≤	less than or equal to
µs	microsecond
%	percent
±	plus or minus
ac	alternating current
AFWAL	Air Force Wright Aeronautical Laboratories
AO	atomic oxygen
APSA	Advanced Photovoltaic Solar Array
C	Celsius
C/NOFS	Communications/Navigation Outage Forecasting System
CHAWS	Charging Hazards and Wake Studies
cm	Centimeter
dc	direct current
DWV	dielectric withstand voltage
emf	Electromotive Force
EMI	Electromagnetic Interference
EMU	Extra-vehicular Mobility Unit (spacesuit)
EOS-AM1	Earth Observing System – Morningside 1 (now Terra)
Eq.	equation
ESA	European Space Agency
EURECA	European Retrievable Carrier
EUV	Extreme Ultraviolet
eV	electron volt
EVA	Extra-vehicular Activity (spacewalk)
F	Farad
FEF	Field Enhancement Factor
FFRDC	Federally Funded Research and Development Center
FPMU	Floating Potential Measurement Unit
FPP	Floating Potential Probe
GEO	Geosynchronous Earth Orbit (35,786 km altitude above mean sea level, 0° latitude)
GRAM	Global Reference Atmospheric Model
GRC	Glenn Research Center
HDBK	handbook
Hz	Hertz
IQSY	International Year of the Quiet Sun

APPROVED FOR PUBLIC RELEASE—DISTRIBUTION IS UNLIMITED

IRI	International Reference Ionosphere
ISO	International Organization for Standardization
ISS	International Space Station
ITAR	International Traffic-in-Arms Regulations
JAXA	Japanese Space Agency
k	Kilo
K	Thousand
KIT	Kyushu Institute of Technology
km	Kilometer
LEO	low Earth orbit (200-2000 km altitude, -50° to +50° latitude, for the purposes of this document)
LeRC	Lewis Research Center (now Glenn Research Center)
m	meter
MEO	medium Earth orbit
MET	Marshall Engineering Thermosphere
ms	millisecond
MSFC	Marshall Space Flight Center
MSIS	Mass Spectrometer Incoherent Scatter
N	Newton
N_2	molecular Nitrogen
NO_2	nitrogen dioxide
NASA	National Aeronautics and Space Administration
NASCAP-2K	NASA/Air Force Spacecraft Charging Analyzer Program
O_2	molecular Oxygen
PAS-6	Space Systems/Loral Commercial Communications Satellite PanAmSat-6
PASP Plus	Photovoltaic Array Space Power Plus Diagnostics
PCU	Plasma Contactor Unit
PIX-II	Plasma Interactions Experiment – II
PMAD	Power Management and Distribution
PMG	Plasma Motor Generator
ProSEDS	Propulsive Small Expendable Deployer System
RC	resistor-capacitor
RCS	Reaction Control System (attitude thrusters)
RTV	Room Temperature Vulcanized-rubber
s	second
S	siemens
S/m	siemens per meter
SAMPIE	Solar Array Module Plasma Interactions Experiment
sec	Second
SI	The International System of Units (commonly known as the Système Internationale)
SPENVIS	Space Environment Information System
SSL	Space Systems/Loral, LLC

STD	standard
STS	Space Transportation System
TSS-1R	Tethered Satellite System – reflight of first mission
U.S.	United States
UV	Ultraviolet
V	Volt
W	Watt
WSF	Wake Shield Facility
XVV	X-axis along Velocity Vector
YVV	Y-axis along Velocity Vector

3.2 Definitions

The following definitions are based on AFWAL-TR-88-4143, Volume 2, Design Guide: Designing and Building High Voltage Power Supplies, Materials Laboratory:

Anode: The electrode through which a direct current enters the liquid, gas, or other discrete part of an electrical circuit; the positively charged pole of an electrochemical cell.

Arc: See Electrostatic Discharge

Breakdown Voltage: The voltage at which the insulation between two conductors fails.

Capacitance (Capacity): That property of a system of conductors and dielectrics that permits the storage of electrical charge in dielectric materials when potential differences exist between the conductors. The value is expressed as the absolute ratio of the stored electrical charge to the potential difference between the conductors. The Standard International (SI) unit is the farad (F).

Capacitor (Condenser): A device whose primary purpose is to introduce capacitance into an electric circuit.

Cathode: The electrode through which an electric current leaves a liquid, gas, or other discrete part of an electric circuit; the negatively charged pole of an electrochemical cell.

Cell: A single unit capable of serving as a direct current (dc) voltage source by transfer of ions in the course of a chemical reaction.

Charge: The fundamental property of certain subatomic particles that determines their interaction with electromagnetic fields. Electric charge is conserved, is either positive or negative, and quantized in integer multiples of the elementary charge, e, approximately equal to 1.602×10^{-19} Coulombs.

Conductance: The reciprocal of resistance. It is the ratio of current passing through a material to the potential difference at its ends. The siemens (S) is the SI unit of electrical conductance. The archaic term for this unit is the mho.

Conductivity: A constitutive property of a material that represents the measure of the material to conduct electrical current in the direction of an externally applied electric field. Conductivity is the reciprocal of bulk resistivity. Its units are siemens per meter (S/m).

Conductor: A material that exhibits a high conductivity that readily permits the flow of electrical current. Practical electrical conductors often comprise a single metallic wire, or a combination of metallic wires not insulated from each other, intended to support the flow of electric current or is similarly conductive and acting as a part of a shield or ground plane structure.

Contaminant: An impurity or foreign substance present in or on a material and affecting one or more properties of the material.

Corona: A non-self-sustaining discharge (sometimes visible) due to ionization of the gas surrounding a conductor where the voltage gradient exceeds a certain critical value for a gaseous medium.

Debye Length: Characteristic distance (λ_p) in a plasma over which the charged particles screen out the electric field by a factor of 0.368 (1/e).

Dielectric: A material that exhibits a high electrical resistivity together with a high electrical breakdown level, such that electrical current does not flow under operational conditions. The term may be synonymous with Insulator.

Dielectric Breakdown: A sudden increase in electric current flow within a dielectric caused by an applied electric field exhibiting a magnitude in excess of the dielectric strength of the material.

Dielectric Strength: The potential gradient at which dielectric breakdown occurs, usually expressed in SI units of volts per meter of thickness (V/m). In some cases, dielectric strength may also be referred to as dielectric withstand voltage (DWV).

Electric Field Intensity: The force per unit charge exerted on a stationary positive charge at a point in space removed from the position of a nearby test charge. It is also known as electric field strength and is a vector quantity. Electric field is defined mathematically as:

$$E = \left[\frac{Q}{4\pi\varepsilon r^2}\right]\hat{r}$$

APPROVED FOR PUBLIC RELEASE—DISTRIBUTION IS UNLIMITED

where r is the distance from the test charge Q, ε is the permittivity of the medium within which the charges reside, and $\hat{r}$ is the unit vector pointing along the radius and away from the positive charge. The SI unit of electric field intensity is newtons per coulomb (N/C) or volts per meter (V/m).

Electrode: An electrical conductor used to make contact with a nonmetallic part of a circuit, e.g. a semiconductor, an electrolyte, a vacuum, or air. Electrodes are commonly used to interface with electrolytic cells, vacuum tubes, gaseous discharge tubes, and free space.

Electromotive Force (emf): Potential difference between two points.

Electron: A stable elementary, negatively charged particle that may be bound to an atom or molecule or free.

Electrostatic Discharge: The sudden transfer of charge between bodies of differing electrostatic potentials. (See Dielectric Breakdown.)

Encapsulating: Enclosing an article in an envelope of plastic or other sealant.

Floating Potential: The potential a spacecraft comes to under current balance ($i_{total} = 0$) with the surrounding plasma. The potential of a surface in a plasma under the current balance condition, i.e. net current to surface equals zero.

Frequency: The number of complete cycles or vibrations per unit of time. The SI unit is the hertz (Hz).

Gradient: The vector derivative of a scalar field producing a vector field with the magnitude of the maximum rate of change of the scalar field and pointing in the direction of the maximum change of the scalar field.

Ground: A common reference plane ("ground" plane) to which electrical circuits are referenced. Typically, the spacecraft structure is the ground plane.

Hollow Cathode: An efficient plasma-emitting device derived from gas flowing through a hot, cylindrical tube cathode and extracted through a closely spaced anode.

Impedance: The total opposition that a circuit offers to the flow of a time varying current. It is equal to the ratio of the time varying voltage (stimulus) to the time varying current (response). It consists of a real part (resistance) and an imaginary part that is circuit dependent. Impedance is measured in ohms.

Insulation: A dielectric material used to prevent leakage of current from a conductor.

APPROVED FOR PUBLIC RELEASE—DISTRIBUTION IS UNLIMITED

Insulation System: All of the materials used to insulate a particular electrical or electronic product.

Insulator: A material that exhibits a high electrical resistivity together with a high electrical breakdown level, such that electrical current does not flow under operational conditions. The term is generally interchangeable with the term dielectric.

Ion: An atom or molecule which has gained or lost one or more electrons relative to its neutral state, giving it a net positive or negative electrical charge,

Ionization: The dissociation of an atom or molecule into positive or negative ions and electrons.

Particulate (space particulate debris): Minute separate particles. The sources of spacecraft particulate debris are Earth, spacecraft, and space environments. Earth particulate is mostly dust, sand, and rocket exhaust. Spacecraft particulate sources can include materials spalled by cosmic dust impacts on materials and the solar array, outgassing products, slip rings, and debris from spacecraft collisions. Space environment particulates consist of residues that form the space plasma, cosmic dust of masses less than one gram, micrometeoroids, and meteoroids.

Paschen Discharge: The result from application of the voltage necessary to initiate a discharge or electric arc between two conductors in a gas as a function of pressure and gap length. The voltage necessary to initiate a Paschen discharge is gas dependent.

Plasma: The fourth state of matter. Plasma typically comprises a gaseous body within which some of all of its constituent atoms are split into ions and free electrons and which exhibits sufficiently low density that considerable charge separation is possible. A plasma is macroscopically neutral and free of any electric fields in its interior, and may be strongly influenced by external electrostatic and electromagnetic fields and forces.

Plasma Arcing: Electrical discharge between two points that results in the creation of a plasma.

Plasma Contactor Unit: A device used on spacecraft to mitigate charging or moderate the floating potential through electron emission.

Plasma Ground: See Plasma Potential.

Plasma Potential: The potential at which neither electrons nor positively charged ions accelerate toward a surface.

Plasma Temperature: The kinetic temperature giving the average thermal energy of an ion or electron in the plasma, often expressed in electron volts (eV). The ion temperature need not be the same as the electron temperature; and, if a magnetic field is present, both can have temperatures that differ parallel and perpendicular to the magnetic field. Typically, the electron temperature is used to characterize a plasma.

Plastic: High polymeric substances, including both natural and synthetic products, but excluding the rubbers, that are capable of flowing under heat and pressure at one time or another.

Polymer: A chemical compound or mixture of compounds formed by polymerization essentially consisting of repeating structural units.

Potential: The work per unit charge required to bring any charge from any point beyond the influence of the electric field to the point of interest.

Power: The time rate of change of energy when work is performed. The SI unit is the watt (W).

Pressure: Force per unit area. Absolute pressure is measured with respect to zero pressure. Gauge pressure is measured with respect to atmospheric pressure.

Pulse: A wave that departs from a first nominal state, attains a second nominal state, and ultimately returns to the first nominal state.

Resistance: Property of a material that determines the current flow produced by a given difference of potential. The real part of the circuit impedance. The SI unit is the ohm.

Resistivity: An intrinsic property of material that quantifies the resistance to charge movement in a material when an electric field is applied. The SI unit of resistivity is ohm-meter but is commonly expressed in ohm-centimeters.

Resistor-Capacitor (RC) Time Constant: Characteristic time obtained by multiplying resistance by capacitance.

Secondary Electron Emission: The release of one or more electrons from a material surface by a primary electron of sufficient energy. Secondary electron emission is a material-dependent property.

Snapover: The phenomenon caused by secondary electron emission that can lead to electron collection on insulating surfaces in an electric field.

Solar Array: Solar cells connected in series and/or parallel to generate power.

Solar Cell: A photovoltaic device used to convert the energy in light to electrical energy.

String Voltage: The voltage of a single series-connected solar array segment.

Sustained Arc: An electrical discharge that lasts longer than 1 ms.

Transient: That part of the change in a variable that disappears during transition from one steady state operating condition to another.

Triple-point: A point in space where insulator, conductor, and plasma all meet.

Voltage: The measure of electrical potential difference between two points. Voltage provides the motive force in response to which electrical current will flow when a conductor is located between the two points. The SI unit is the volt (V).

Wire: A metallic conductor generally of round, square, or rectangular cross-section that can be either bare or insulated.

4. ENVIRONMENTS

4.1 The Ambient Environment

4.1.1 The Neutral Atmosphere

The dominant environment between 100 and 1000 km is the neutral atmosphere as seen in Figure 1, Composite International Year of the Quiet Sun (IQSY) Daytime Ionosphere and Atmosphere Based on Mass Spectrometer Measurements, which depicts the neutral densities for quiet solar conditions. In this essentially collision-less regime, the gases are in hydrostatic equilibrium. Below about 100 km, where the atmosphere is homogeneous, the composition is approximately 80 percent N_2 and 18 percent O_2 with trace amounts of NO_2, argon, and other gases. Above 100 km, atomic oxygen is the dominant constituent as a result of photo-dissociation of molecular oxygen. At roughly 500 km altitude, the neutral number density varies from 2×10^6 to 3×10^8 cm^{-3}, depending on solar activity and position in the orbit. The kinetic temperature of the gas lies between 500K and 2000K while the ambient pressure lies in the range of 1×10^{-10} Torr to 5×10^{-8} Torr. Above the 800 – 1000 km region, the atmosphere is largely atomic hydrogen.

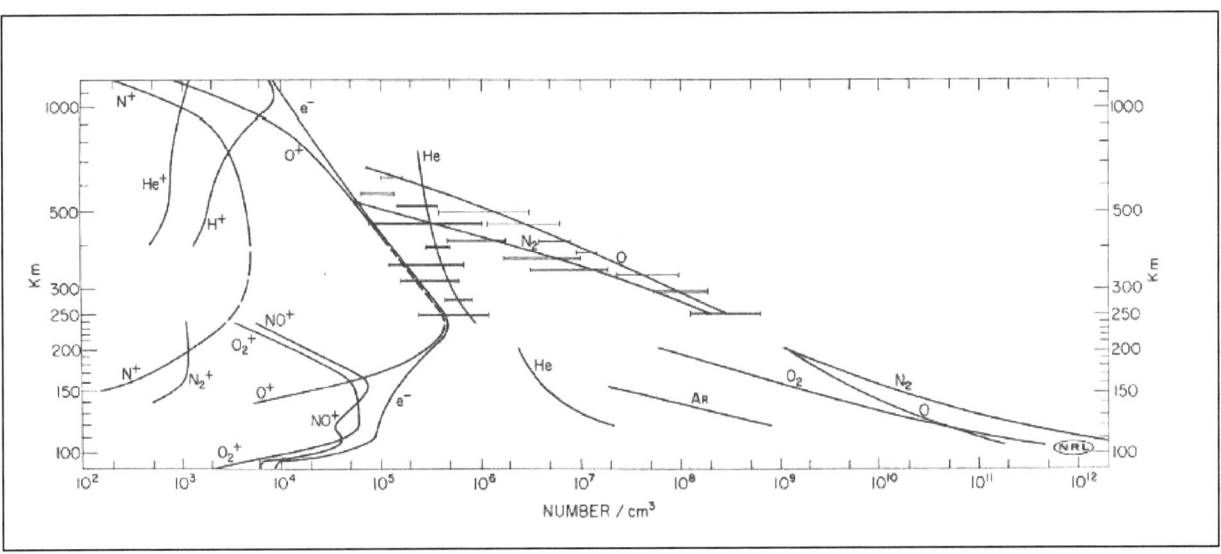

Figure 1—Composite International Year of the Quiet Sun (IQSY) Daytime Ionosphere and Atmosphere Based on Mass Spectrometer Measurements
(Johnson 1969, reprinted with permission of MIT Press)

In Figure 1, ion and neutral composition distributions below 250 km are from two daytime rocket measurements above White Sands, New Mexico. (The helium distribution is from a nighttime measurement.) Distributions above 250 km are from the Electron 2 satellite results of Istomin (1966) and Explorer 17 of Reber and Nicolet (1965). The bars show the variations of the respective neutral constituents within the atmosphere due to latitudinal, diurnal, and storm effects.

The neutral gas environment has been well explored and quantified. Empirical models based on *in-situ* neutral composition and satellite drag measurements have evolved over the years into reliable predictors of the average composition and thermal structure of the thermosphere. The most prominent models are the Mass Spectrometer Incoherent Scatter (MSIS-86) model (Hedin, 1987; Prag, 1983) based on *in-situ* satellite observations of neutral concentrations, the Marshall Space Flight Center (MSFC) version of the Jacchia model derived from satellite drag measurements, the Marshall Engineering Thermosphere (MET-2007) ("Computational procedure," 1970; Hickey, 1988), and the U.S. Standard Atmosphere ("U.S. standard," 1976; King, 1978). All of these models are included as options in the more recent Earth Global Reference Atmospheric Model (GRAM) described in Leslie and Justus (2011). These models provide good estimates of the thermosphere environment as functions of altitude, longitude, latitude, local time, magnetic activity, and solar activity and are continually updated as new information becomes available. Additional references for atmospheric models are BSR/AIAA G-003C-2010, American National Standard, Guide to Reference and Standard Atmosphere Models, and ISO TR 11225 (2012), Space Environment (Natural and Artificial) – Guide to Reference and Standard Atmosphere Models.

APPROVED FOR PUBLIC RELEASE—DISTRIBUTION IS UNLIMITED

4.1.2 The Plasma Environment

On the sunlit hemisphere of the Earth, ultraviolet (UV) and extreme-ultraviolet (EUV) radiation penetrates the atmosphere, ionizing and exciting the neutral constituents, forming a region in which electrons and ions are present. The ionosphere is a highly dynamic plasma whose properties vary with altitude, latitude, time of day, and sunspot cycle. The variability with latitude, known since the 1920s, is so dramatic that the ionosphere is conventionally divided into three distinct regions: high-latitude (> 50 deg), mid-latitude (20 < deg < 50), and low-latitude (< 20 deg). Local geomagnetic disturbances can cause dramatic variations in the ionospheric properties that are difficult to predict and can range in duration from hours to weeks. Despite these complications, the broad features of the ionosphere can be described with simple models. A widely recognized model is the global International Reference Ionosphere (present version at IRI-2012) which provides estimates of plasma concentrations, composition, and temperatures as a function of altitude, time, location, and solar activity condition. (IRI software and support can be obtained from *http://iri.gsfc.nasa.gov*)

The low latitude ionosphere has recently been explored by the Air Force-developed Communications/Navigation Outage Forecasting System (C/NOFS) mission to investigate and forecast scintillations that drastically affect communication [de la Beaujardiere, 2004]. The mid-latitude region has been the most explored and comprises the bulk of data for the models in the IRI. In this document, spacecraft design and its interaction with the local environment is not treated for the high-latitude region. See the latest revision of NASA-HDBK-4002 for guidelines on mitigating charging for all other orbits, i.e., high latitude LEO, medium Earth orbit (MEO), and GEO.

Variation with altitude is perhaps the most important parameter for the spacecraft designer. This pronounced vertical structure is not simply a matter of height variation but reflects basic physical processes that differ in the respective regions. Three processes, in particular, are responsible: (1) the sun's energy is deposited at various altitudes because of the absorption characteristics of the atmosphere, (2) the physics of recombination depends on density and therefore on altitude, and (3) composition of the atmosphere changes with altitudes.

The lower limit of the ionosphere is somewhat arbitrary since plasma production falls off continuously with decreasing height. Historically, the ionosphere has been assumed to begin at about 50 km from the surface because at this altitude the plasma density becomes sufficient to noticeably affect radio wave propagation. There is no distinct upper limit for the ionosphere, but 2000 km is generally used for most practical applications. In this document, spacecraft design for altitudes above 1000 km is not specified because radiation issues become the primary design driver.

Four layers describe the vertical structure of the ionosphere. In order of increasing altitude and increasing plasma density, these layers are designated as D, E, F1, and F2 regions. The properties of each layer are summarized in Table 1, Nominal Properties of Ionospheric Layers.

APPROVED FOR PUBLIC RELEASE—DISTRIBUTION IS UNLIMITED

Table 1—Nominal Properties of Ionospheric Layers

Region	Nominal Height of Peak Density (km)	Plasma Density at Noon (cm^{-3})	Plasma Density at Midnight (cm^{-3})	Dominant Ion
D	90	~1.5 x 10^4	Vanishes	O_2^-, O_2^+, NO^+
E	110	~1.5 x 10^5	~1 x 10^4	O_2^+
F1	200	~2.5 x 10^5	Vanishes	O^+
F2	300	~1.0 x 10^6	~1.0 x 10^5	O^+

For altitudes higher than that of the peak density in the F2 layer, the electron density decreases monotonically out to several Earth-radii, the distance dependent on geomagnetic activity and convection electric field (Chappell, 1972; Bailey, et al., 1997; Gallagher, et al., 2000). For altitudes up to and including the F2 density peak, electron and ion thermal energies are in the range of 0.1 eV to 0.2 eV, corresponding to kinetic temperatures of 1200K to 2400K. The temperature increases monotonically beyond the F2 density peak and reaches several thousand eV in geosynchronous orbits (~ 42,164 km) (Garrett, et al., 1978). The F2 layer is important for spacecraft operations. It is in this layer that the International Space Station (ISS) and most LEO spacecraft orbit the Earth.

Ionospheric plasma distributions within the F-region have been extensively explored since the advent of bottom-side sounders, long before *in-situ* satellite observations were made. As a result, the general morphology of the F-region and some of its more prominent individual features are well understood. While there are detailed features such as localized troughs, localized heating, and short temporal variations that are difficult to model, the overall global structure of the ionosphere is now well understood; and excellent ionospheric models exist for estimating and quantifying plasma distributions. A widely used model is the IRI-2012 (see above).

4.2 The Spacecraft-Induced Environment

Spacecraft-induced environments can take many forms, including but not limited to, neutral gases, ionized gases (plasmas), condensable gases, particulates, and radiation. In many cases, these environments can overwhelm the natural environment and can lead to undesirable interactions. These types of environments are treated separately in the following sections.

4.2.1 Neutral Gas

Cold gas thrusters and Reaction Control System (RCS) thrusters can significantly increase the localized neutral pressure. This increase can result in Paschen discharges in the presence of exposed high-voltage conductors. In general, if the local neutral pressure is more than a millitorr and less than a few torr, high-voltage electrical breakdowns can occur (see Figure 2, Voltage Breakdown of Pure Gases as a Function of Pressure Times Spacing). On the Solar Array Module Plasma Interactions Experiment (SAMPIE) (Ferguson and Hillard, 1997) that was carried in the Shuttle payload bay, a local gas vent had to be moved to prevent Paschen discharge.

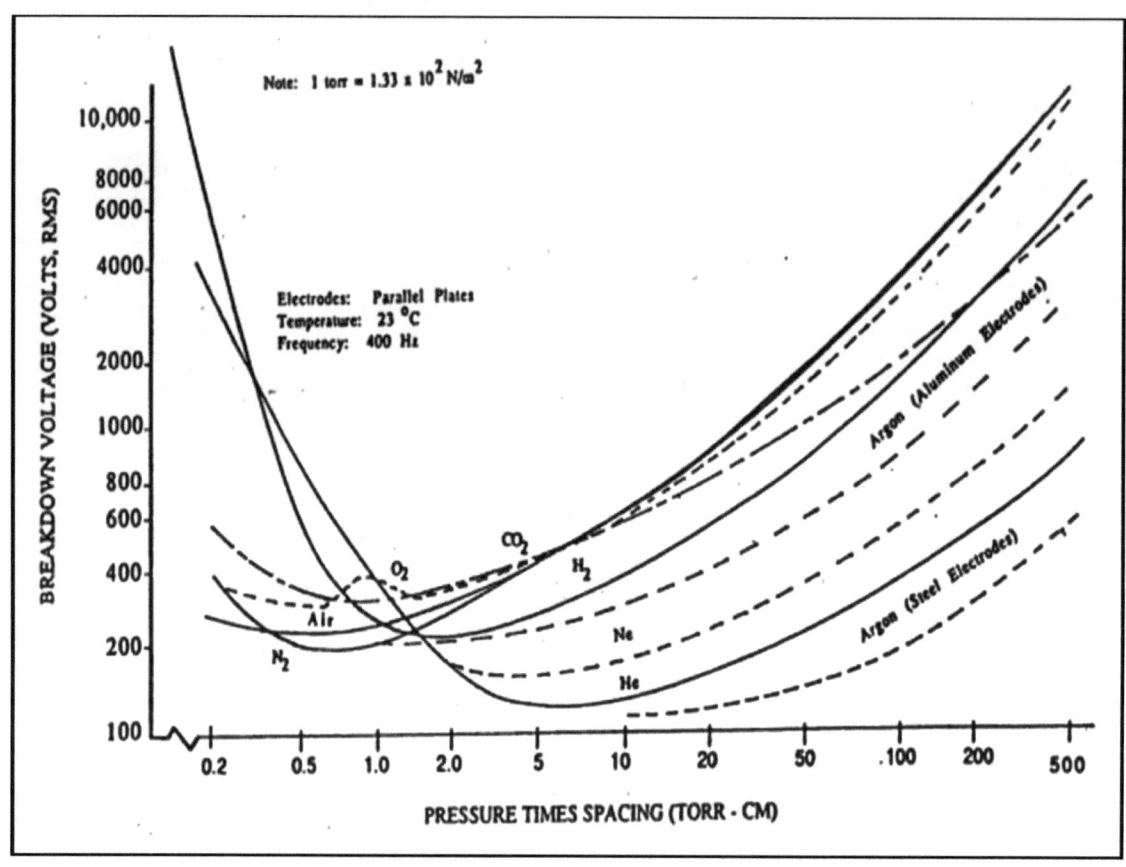

Figure 2—Voltage Breakdown of Pure Gases as a Function of Pressure Times Spacing (Paschen curves for different gases, from Dunbar, 1988)

4.2.2 Ionized Gas/Plasma

Ionized gases can be emitted by plasma sources such as hollow cathode plasma contactors or from neutral gas sources at high positive potentials. Locally, the plasma density can be greater than the ambient plasma density, and similar plasma interactions can occur with high-voltage components. On ISS, the plasma contactor units (PCUs), when operating, produce a xenon plasma plume that is much greater in density than the ambient plasma. It has been estimated that the plume of the ISS PCU extends several 10s of meters along the magnetic field and can easily impact surfaces of the ISS solar arrays (Gabdullin, et al., 2007). Arcing and current collection from such a plasma could occur in much the same way as with an ambient plasma, implying that solar arrays and other active sites should be kept out of induced plasma plumes.

4.2.3 Condensable Gas

Condensable gases are effluents that can condense on cold components and contaminate their surfaces. Oils and water vapor are two major condensable gases that can influence the interactions of spacecraft surfaces. Many hydrocarbon oils, however, will break down upon interaction with the LEO atomic oxygen environment on ram-facing surfaces but can build up on wake surfaces. Silicones will react with the LEO atomic oxygen to form a silicate layer on surfaces.

Water vapor released on the night side can condense on insulating surfaces of solar arrays, etc., and can participate in arcing when the arrays become active in sunlight. It has been shown in laboratory testing that solar arrays that have undergone vacuum bakeout for seven days lose the water vapor contamination that is an important contributor to low-voltage (-100 to -300 V) arcing (Vayner, et al., 2002). In LEO, however, a cold cycle is about 1/3 of every orbit. Even very well baked-out systems can have effluents evolved during the day side re-condense on the night side of the orbit. Thin layers of condensed contaminants can concentrate electric fields above high-voltage conductors, even to the point where they undergo dielectric breakdown.

4.2.4 Particulates

Particulates can be emitted or shaken from surfaces, but can also result from arcing or sputtering from spacecraft surfaces. Particulates can transfer small amounts of charge from one surface to another, but their major effect is in changing the characteristics of the surfaces to which they adhere. For instance, an insulating particle on a conductor that is at a high potential can concentrate the electric field structure locally, possibly leading to a reduced arcing voltage threshold.

4.2.5 Radiation

High energy charged particles, e.g. electrons, can imbed deep within dielectrics where they can accumulate over extended time periods until the dielectric breaks down under the induced electric field. In the natural environment, this effect will mainly happen in the auroral zones, radiation belts, and above the South Atlantic Anomaly, and thus are not usually important in the environment for which this NASA Technical Handbook is applicable. However, radiation produced on or within a spacecraft can be important regardless of orbital position. Satellites using radioactive power sources must be designed to prevent this "deep-dielectric" charging, which is different from the typical "surface" spacecraft charging.

NASA-HDBK-4006A

5. PLASMA INTERACTIONS

5.1 Poisson Equation

When energized conductors are exposed to plasma, positive surfaces collect electrons and negative surfaces collect ions. The Poisson equation governs potential distributions which determine charge movement. The Poisson Equation is as follows:

$$\nabla^2 \phi = -\rho/\varepsilon_0 \qquad \text{(Eq. 1)}$$

where ϕ is the potential, ρ is the charge density, and ε_0 is the permittivity of free space.

Electrons, which are much lighter and more mobile than ions, are more readily collected. Surfaces charge to whatever potential, relative to the surrounding plasma, necessary such that the net current equals zero in equilibrium as shown in Eq. 2:

$$I_{ion} + I_{electron} + I_{photoelectrons} + I_{secondary\ electrons} + I_{other\ sources} = 0 \qquad \text{(Eq. 2)}$$

where I_{ion} = ion current, $I_{electron}$ = electron current, $I_{photoelectron}$ = photoelectron current, $I_{secondary\ electron}$ = secondary electron current, and $I_{other\ sources}$ = current from all other sources. The first four terms are usually dominant. Current closure in space requires the ambient plasma as part of the conducting path. In the case of LEO, this is the ionosphere plasma.

Predicting the potential for complex spacecraft surfaces in space can be achieved through either numerical computational models or laboratory plasma testing or a combination of both. The resulting interactions can be summarized as follows:

    a. Surfaces that are more negative than approximately -100 V with respect to their surroundings are subject to arcing. These arcs can be either plasma arcs or arcs to adjacent conductors. They are usually a momentary discharge of accumulated energy, lasting only microseconds to milliseconds, depending on arc type but, under some conditions, can be sustained. The necessary conditions for the arc to be sustained are for the current and voltage to be maintained above threshold values for durations lasting milliseconds to seconds. Plasma arc thresholds can be as low as -75 V under some conditions. More information about arcing can be found in ISO 11221, Space systems – Space solar panels – Spacecraft charging induced electrostatic discharge test methods, and Bodeau, 2012 and 2014.

    b. Surfaces that are more negative than approximately -40 V are subject to ion bombardment and sputtering (Rutledge, et al., 1992). Since the dominant ion in LEO is atomic oxygen, care must be taken to avoid chemical attack as well.

    c. Solar array surfaces with exposed conductors that are positive with respect to the ambient plasma will collect electrons from the ambient plasma. This phenomenon is referred to as "parasitic current collection" and can result in measurable solar array current (and power) loss of a few percent or more for solar arrays operating at higher voltages. Likewise, solar arrays with large amounts of positive exposed metal conductors may also suffer increased parasitic current collection (Goebel, et al., 2014).

    d. If the solar array string return conductors are grounded to the spacecraft ("negative grounding"), the spacecraft chassis can float negative with respect to the ionosphere. The spacecraft potential can become as negative with respect to the ionosphere as 90% of power system voltage if sufficient conducting area is not available to collect enough ions to balance electron collection (see section 5.2.1.2). For systems with very large areas of high-voltage surfaces such as the ISS, this effect is large, requiring a plasma contactor to mitigate it. Note that when the ISS has its plasma contactor (grounded to the structure) operating, the ISS structure floating potential at all points on the station is held to ±40 V with respect to the local space plasma potential. Current collection of the solar arrays (from the plasma) is exacerbated because the arrays will be held at positive potentials with respect to the surrounding plasma.

    e. In some situations, electron current collection will dramatically increase, a situation referred to as "snapover." See section 5.2.1.3.2 for a more detailed discussion of snapover.

5.1.1 Plasma Sheath and Debye Length

A positively charged electrode immersed in a plasma will collect electrons and repel ions. The volume surrounding the electrode that is influenced by the charged surface is called the "sheath." Outside of the sheath, the plasma particles are not affected by the charge (or voltage) on the electrode surface. The extent of the sheath is dependent on the electrode voltage, plasma density, and electron temperature. A distance measure of the sheath is given by the Debye length (λ_d) the formula for which is shown in Eq. 3:

$$\lambda_d = \left(kT_e / 4\pi n e^2 \right)^{1/2} = 743 \left(T_e / n \right)^{1/2} \text{ cm} \tag{Eq. 3}$$

where k= Boltzmann's constant, T_e = electron temperature in eV, n = plasma density in cm^{-3}, and e = electron charge. (Chen, 1984)

In general, the sheath extends from the electrode surface a few (≤ 10) Debye lengths. For higher voltages, i.e., much larger than the electron temperature ($> 10T_e$), some electrons may orbit around the electrode and escape from the sheath. The collected or trapped electrons are said to be orbit-limited and are affected in a complex manner by the radius of the electrode, the electrode voltage, the electron temperature, and electron density.

5.2 Exposed Biased Conductors

It is good practice to minimize the amount of exposed biased conductors on spacecraft. Exposed biased conductors at sufficient voltage ($|V| > 55$ V) that do not exhibit corona or Paschen breakdown in a neutral gas can readily do so if the environment contains a significant ionized component. High-voltage surfaces by design, e.g., solar cell interconnects, can be exposed to the space plasma and driven to high potentials because of current collection from the space plasma. The resulting equilibrium potentials on spacecraft surfaces may cause the following deleterious effects: large floating potential variations, parasitic power drain, sputtering, and arcing.

5.2.1 Current Collection

An extensive resource of papers for this topic can be found in Current Collection from Space Plasmas (Singh, et al., 1990).

5.2.1.1 Current Collection by Structures

5.2.1.1.1 Electron Collection

For electron thermal temperatures in LEO, the ambient electrons are moving at speeds greatly in excess of the orbital velocity. The ISS, for example, has a spacecraft velocity of 7.7 km/s, compared to the typical electron thermal velocity of 180 km/s. Thus, electrons can be collected on any conducting surface not charged to a potential that is more than a few times the ambient potential negative. In general, electron collection is well described by probe theory. See, for example, Chen, 1965. For large surfaces, collection is best described by thin sheath probe theory. For surfaces whose characteristic dimension is less than the Debye length, orbit-limited theory can be used.

Electron current collected from a plasma can be described by the equation

$$I_e = J_0 A_s \qquad (Eq.\ 4)$$

where I_e is the electron current, J_0 is the electron thermal flux, and A_s is the effective surface area for electron collection. The collection area is a function of collecting conductor surface geometry and adjacent surface electrical properties and geometry. In relatively simple geometries (e.g., planar surfaces), the collecting area can be either the plasma sheath area or the area of a sphere with the limiting orbit radius. The electron thermal flux, J_0, is given by

$$J_0 = (ne)(kT_e/2\pi m_e)^{1/2} = 2.5 \times 10^{-14}\, n T_e^{1/2}\ \text{Amps/m}^2 \qquad (Eq.\ 5)$$

where n is the electron density per m^3, and T_e is the electron temperature in eV. For example, at the F2 peak in the ionosphere where the density is $\sim 1 \times 10^{12}/m^3$ and temperature is ~ 0.15 eV, the electron thermal flux is ~ 9.7 mA/m^2.

Electron current collection by wires is important in the case of electrodynamic tethers or when structures such as self-extending masts with wire braces are used. For instance, when the ISS is in nominal YVV (Y-axis in the velocity vector) flight mode, all solar array mast wires and flex blanket guy wires will collect ions. However, when in XVV (X-axis in the velocity vector) flight mode, half of the solar array masts and flex blanket guy wires will be electron collecting, and half will be ion collecting. The array wing, positive with respect to the plasma because of magnetic induction effects (see Eq. 6), acts as an electron collector, and becomes essentially grounded to the surrounding plasma. The electromotive force induced in a conductor in relative motion to a magnetic field is given by

$$\mathbf{emf} = \vec{v} \times \vec{B} \cdot \vec{l} \qquad \text{(Eq. 6)}$$

where $\vec{v}$ is the velocity, $\vec{B}$ is the magnetic field, and $\vec{l}$ is the length of the conductor (tether, structure, etc.).

An electrodynamic tether is a long wire orbiting in the Earth's magnetic field that uses the electric field generated by its motion (see Eq. 6) to produce power or propulsion. This concept was proved on orbit by the Plasma Motor Generator (PMG) experiment, where both modes of operation were produced by emitting electrons (by means of plasma contactors) either at the top or bottom of a 500-meter tether to produce power (electron emission at the bottom) or propulsion (electron emission at the top). The maximum $\vec{v} \times \vec{B}$ on a structure in LEO is about 0.25 volts per meter (Grossi, 1995). Propulsion is derived from the body force ($\vec{F}$) on the tether due to

$$\vec{F} = \vec{I} \times \vec{B} \qquad \text{(Eq. 7)}$$

where $\vec{I}$ = current in tether and $\vec{B}$ is the magnetic field.

In the case of the Tethered Satellite System TSS-1R mission, the final deployed tether length of 19.7 km produced nearly 3500 V potential between the ends of the tether. The satellite at its upper end (the positive end) collected electrons, and an electron gun at the lower end emitted electrons to complete the circuit (Dobrowlny and Stone, 1994). The electron collection by the satellite proved more efficient than pre-mission prediction (Stone, et al., 1999). Following the tether separation from the Space Shuttle Orbiter after an arc in the deployer (see section 6.2.1), the electron collection by the satellite was ~ 1.1 A for approximately 70 sec (Gilchrist, et al., 1998). The end of the severed tether apparently turned into an efficient plasma contactor (see section 7.2.2) to sustain this current (Gilchrist, et al., 1998).

Electron collection in LEO is also affected by the vehicle plasma wake. Plasma wakes are discussed in section 5.2.1.5.

Insulating structure surfaces reach equilibrium potential with the LEO plasma of only a few volts negative and do not thereafter collect current (Vaughn, 2003).

5.2.1.1.2 Ion Collection

Ions in LEO, due to their large mass and low thermal velocity, are mainly collected by spacecraft ram surfaces. For the ISS, the spacecraft velocity is 7.7 km/s, whereas a typical ion thermal velocity is only 1.1 km/s. Since many conducting parts of a structure are far greater in dimension than the plasma sheath, the effective flux of ions to their surfaces is essentially equal to the ram flux of ions on their front-facing surfaces. The ion current is the integrated current density (flux) over an area expressed by

$$I_i = J_i A_{ram} \tag{Eq. 8}$$

where I_i is the ion current, J_i, is the ram ion current density (flux), and A_{ram} is the ram surface area for ion collection.

The ion current density, J_i, can be written as

$$J_i = qnV \tag{Eq. 9}$$

where q = e for electrons and Ze for ions, Z = charge state, n = density and V = spacecraft orbital velocity ~ 8 km/s. At the daytime F2 peak, the density is ~ 1×10^{12} m^{-3}. The LEO ion current density is ~ 1.3 mA/m^2.

5.2.1.2 Floating Potential: The Current Balance Condition

In the weakly ionized low-density plasma found in LEO, current collection is completely described by Poisson's equation (Eq. 1). Electrons are easily attracted to positive surfaces while collection of ions to negative surfaces is limited due to their much larger mass (mostly ram ion collection). In equilibrium, net current collection must be zero (Eq. 2); that is, the net current of each polarity is identical. Focusing on the plasma currents, and ignoring secondary and photo emission, Eq. 2 can be reduced to

$$I_i = I_e \tag{Eq. 10}$$

In other words, the equilibrium floating potential occurs when the ion current, I_i, is equal to the electron current, I_e. (Note the polarity of the electron current is negative.)

Recall from Eq. 4 and Eq. 8, current can be expressed as a current density multiplied by an area, I = JA. In the case where the collection area is the same such as a ram facing area, Eq. 10 can be further reduced to

$$J_i = J_e \tag{Eq. 11}$$

where J_i is the ion current density, and J_e is the electron current density.

To illustrate how the current balance condition can be used to calculate a floating potential, we consider the example provided by Tribble, 1995 (entitled "A Biased Object in LEO"). In his example, Tribble describes a conventional and simplified LEO solar array comprised of a number of photovoltaic cells connected in series via exposed metal interconnects. The series connection of cells forms what is known as a string. Each metal interconnect is capable of collecting some amount of current from the LEO plasma. The relative bias of each interconnect will depend on its position within the string. For example, an interconnect at the beginning of a string will only be biased at the output of one cell, ~1 volt, whereas an interconnect at the end of the string will be biased at nearly the full solar array voltage, perhaps as much as 100 volts. In Tribble's example, he assumes the array is ram facing, so ions may be collected as well as electrons. Tribble also notes that ions will be reflected (not collected) if the interconnect bias is more positive than the ion impact energy (ϕ_i). Likewise, electrons will not be collected if the interconnect bias is more negative than their impact energy (ϕ_e).

From Tribble Eq. 4.43, the ion current density (J_i) and electron current density (J_e) are defined as

$$J_i = en_0 v_i \left(\frac{fV_a - \varphi_i}{V_a}\right) \tag{Eq. 12}$$

$$J_e = en_0 v_{e,th} \left[\frac{(1-f)V_a - \varphi_e}{V_a}\right] \tag{Eq. 13}$$

where e is the electron charge, n_o is the plasma density, v_i is the ion velocity, v_e is the electron velocity, V_a is the solar array voltage, and f is the fraction of the solar array that floats negatively (relative to the plasma).

As shown in Eq. 11, for the case of equal collection areas, the floating potential occurs when the ion current density is equal to the electron current density $J_i = J_e$. Substituting the current density equations provided by Tribble into Eq. 11, one gets

$$en_0 v_i \left(\frac{fV_a - \varphi_i}{V_a}\right) = en_0 v_{e,th} \left[\frac{(1-f)V_a - \varphi_e}{V_a}\right] \tag{Eq. 14}$$

From Eq. 9 for nominal LEO conditions, it was shown $J_i = en_0 v_i = 1.3 \; mA/m^2$

From Eq. 5 for nominal LEO conditions, it was shown $J_e = en_0 v_{e,th} = 9.7 \; mA/m^2$

Substituting these current density values into Eq. 14, one gets

$$1.3 \left(\frac{fV_a - \varphi_i}{V_a}\right) = 9.7 \left[\frac{(1-f)V_a - \varphi_e}{V_a}\right] \tag{Eq. 15}$$

Rearranging terms, one finds

$$7.5 \, \varphi_e - \varphi_i = 7.5 \, V_a [1 - 1.3 f] \tag{Eq. 16}$$

Let φ_i = 5 eV, φ_e = 0.15 eV, which are typical impact energies for LEO plasma conditions. (Where the ion energy is the ram energy due to the spacecraft velocity, and the electron energy is related to the random thermal motion of the electrons.)

Solving for f (the fraction of the array that floats negative), one arrives at

$$f = \frac{1}{1.13}\left[1 + \frac{0.52}{V_a}\right] \qquad \text{(Eq. 17)}$$

For many years, a typical satellite solar array voltage was 28 volts, i.e., V_a = 28V. In this case, f = 0.90 or 90% of the array collection area floats to negative potentials.

The quest for high power satellites has moved solar array voltages higher. The ISS operates one of the highest solar array voltages at 160 volts. Applying Eq. 17 to ISS, V_a=160V, then f = 0.89 ~ 0.9. Again, the fraction of the solar array collection area that floats negative is effectively 90%.

Tribble's example shows that floating potentials of biased objects in LEO tend towards negative values as a result of the relatively small ion current density compared to the electron current density. To achieve current balance (ion current = electron current), a significant portion of the collecting area must float to negative voltages (with respect to the plasma) to accumulate a sufficient amount of the slow-moving ions. In the examples shown for both low- and high-voltage arrays, approximately 90% of the collecting area was driven to negative floating potentials. Thus, the rule of thumb for estimating the floating potential of a biased object in LEO is a negative potential approximately equal to 90% of the bias voltage.

In the case of ISS, the power system consists of solar arrays wired in a series-parallel arrangement to give a 160-volt system. Since the main structure of ISS is "grounded" to the negative end of the array string, the entire ISS was predicted to "float" more than 140 V negative with respect to the LEO plasma (ionosphere). Such potentials are beyond the dielectric strength of the anodized coatings on the ISS aluminum structure and would lead to arcing into the space plasma and eventual destruction of the anodized coatings which serve as passive thermal control for the ISS. To eliminate the possibility of arcing of the anodized coating, an active plasma contactor, a xenon hollow cathode discharge unit, was added to ISS to effectively ground to the ionosphere. However, the ISS proved to have an interesting interaction with the LEO space plasma. Although the welded-through design of ISS solar array interconnects was intended to inhibit electron collection, the low energy LEO electrons were able to access the cell edges leading to electron current collection. When *in situ* ISS floating potential measurements were ultimately made, it was determined that 25 – 35 m^2 of ram ion collection area existed on ISS (depending on its flight attitude). This ram ion collection significantly offset the electron collection, thereby keeping ISS floating potentials below that which would threaten the ISS thermal control coating. The ISS plasma contactor is only used for crew extravehicular activity (EVA) events (Reddell, et al., 2006; Wright, et al., 2008; Craven, et al., 2009; Minow, et al., 2010).

For conducting surfaces that are covered with insulators, some elapsed time could be necessary for the steady state potential situation to be reached. The surfaces will charge until no further charge collection is necessary in equilibrium, and this is tantamount to charging up a capacitor with plate separation equal to the insulator thickness. Ion charging times in LEO can be considerable for typical anodized aluminum thicknesses. It is estimated, for instance, that in the daytime ionosphere, ISS surfaces will take between 3 - 4 seconds to fully charge due to ram ion current, whereas on the morning terminator where the ionospheric ion density is at its lowest, charging times of 40 seconds or more can occur (Carruth, et al., 1992).

Notice that for most purposes, the collected ion current depends only on the plasma (ion) density, whereas the electron current depends on both the electron density and the electron temperature. To first order, then, when there is a current balance condition determining the floating potential, only changes in the electron temperature will cause changes in the floating potential.

Insulated ram surfaces will float at a potential such that the ram ion and thermal electron currents are equal, which equates to only a few volts negative at the most.

5.2.1.3 Current Collection by Solar Arrays

5.2.1.3.1 Electron Collection

Electrons can be collected on positively charged cell interconnects, exposed wire traces, or exposed cell edges of the solar arrays. Since solar arrays generate a voltage across each string, some of the array components will be at different voltages. If a solar array string has 400 silicon solar cells in series, with each cell producing 0.5 V, one end of the string will be about 200 V more positive than the other. The total electron current collected will be the integral across the array of the collection from each cell at its respective potential relative to the plasma potential. This depends, of course, on the system grounding configuration, i.e., positive ground or negative ground. Wherever the system floats with respect to the ambient plasma, only the cells and traces with positive potentials will collect electrons.

If the array's exposed conductors are partially hidden from the ambient plasma (such as being underneath overhanging coverslides or between closely spaced solar cells), the coverslides can change the electron collection greatly. Early studies concerning ISS charging and array construction showed that a coverslide with an overhang at least as big as the cell-plus-adhesive thickness will reduce electron collection at the cell edge by a few orders of magnitude (Ferguson, 1991; Ferguson, et al., 1990). One way of thinking about this reduced electron collection is that it becomes difficult or impossible for thermal electrons to reach the cell edges, due to their trajectories. For such solar arrays, it is often the case that for lower ambient electron temperature, the electron collection will be greater, since a greater number of ambient electrons have trajectories that can impact the cell edge. This is the case for the ISS arrays, where the largest amount of electron collection, and thus the most negative charging, occurs when the ambient electron temperature is lowest (Mandell, et al., 2003).

The solar array itself can provide a wake to block its own electron collection. For a non-gimbaling array in equatorial LEO, the electron collection will be at a maximum near sunrise, and will shut off about noon when the array goes into its own wake. Of course, at night when the plasma density is much lower and the array is not generating voltage, electron collection will be minimal. Thus, for non-tracking solar arrays, electron collection in LEO is only important, and can only lead to a great deal of system charging, for about 1/3 of each orbit (the morning side). An important exception to this is the ISS solar arrays, which have bifacial power production and often are locked in position (not sun tracking). This allows for back-side sun pointing, producing full string voltage with plasma ram densities on the solar array front and allowing for another period of high electron collection.

5.2.1.3.2 Snapover

The term "snapover" is associated with dramatically enhanced electron current collection. Snapover occurs during high-voltage operation of a positively biased exposed conductor next to a dielectric surface. A solar array, with exposed interconnects between insulating coverglass surfaces, is a common example of an arrangement that can be subject to snapover. The occurrence and magnitude of snapover effects are directly dependent on the physical layout of the conductor-dielectric interfaces, as well as the dielectric material properties.

Snapover was first observed in laboratory tests in the mid-1970s (Stevens, et al., 1978). Researchers testing positively biased solar array samples exposed to LEO plasma conditions found that as they increased the bias voltage a threshold was reached, the onset voltage, where there was a dramatic increase in current collected. When estimating the area required to collect such a large current, it appeared the whole sample surface was somehow collecting electrons—despite the fact that much of the sample was made up of dielectric coverglass materials. In the late 1970s, the Plasma Interaction Experiment (PIX) missions showed snapover does occur on-orbit (Grier and Stevens, 1978; Grier, 1985). The PASP-Plus mission in the mid-1990s reaffirmed the occurrence of snapover in spaceflight conditions (Davis, et al., 1998).

Excessive current collection as a result of snapover can dramatically impact a solar array power system, as power generated by the array is wasted on driving current through the plasma rather than being applied strictly to the intended spacecraft loads. In other words, snapover exacerbates parasitic power drain (see section 5.2.1.3.4).

Two factors appear common to all theories related to snapover: (1) the generation of a large number of secondary electrons due to LEO plasma electrons being accelerated to positively biased surfaces; and (2) the formation of an extended plasma sheath which inflates the effective collection area. A schematically illustrated description of snapover, as well as relevant laboratory test data, is provided by Iwasa, et al. (2006).

An example to help understand snapover is the pinhole scenario, described in Stillwell, et al., 1985. Suppose a flat conducting plate is covered with an insulator, and in this insulation there is a pinhole. If the plate is biased positive by a power supply and placed in plasma, it will collect electrons. For low voltages, electron collection will be linear with bias voltage. Although the insulating surface cannot collect charge, it nevertheless is the source of an increasing electric field as charges accumulate on its surface. This field results in ion bombardment of the insulator and subsequent secondary electron emission. The result is a rapidly growing sheath that collects charge and funnels it effectively to the pinhole. What is observed then is this: As voltage is increased from zero, current is collected linearly. At some point, current collection increases exponentially and finally saturates at a current level that is approximately the same as if the entire plate were conducting. On a solar array, the interconnects, wire traces, or cell edges act like pinholes; they are the conductors to which the current is funneled. The solar cell substrate and/or coverslides act like the insulator in the above example; they are the dielectric that furnishes the secondary electrons, and in snapover conditions they act as a current-collecting plate.

Avoiding snapover is an important design issue, particularly with arrays designed to operate at voltages above 100 V. Strategies include insulating all surfaces, where practical, and choosing insulators with low secondary electron emission yields. While simply insulating all conducting surfaces provides initial protection, cracks or pinholes are difficult to avoid when materials must withstand years of exposure to harsh space conditions. It should be noted that pinholes in high-voltage insulation usually expand as the large current density funneled through them destroys additional material. On the other hand, experience has shown that cracks or pinholes, if much smaller than the Debye length in the plasma, do not snap over. $\lambda_D = 743(T_e/n)^{1/2}$, in cm, where T_e is the electron temperature in eV, and n is the electron density in cm^{-3}. (See Eq. 3 in section 5.1.1. For LEO conditions, λ_D can be as small as 0.1 cm.)

As an example of a snapover-like effect on real solar arrays, consider the data in Figure 3, Electron Current vs. Bias for Three Solar Array Blanket Materials. The Advanced Photovoltaic Solar Array (APSA) was a very lightweight design proposed for widespread use in the early 1990s. Originally designed for deployment in GEO, the blanket material was carbon-loaded Kapton®, which had sufficient conductivity to avoid the differential charging that is a common problem in that environment. Proposals to adapt APSA technology to LEO recognized that atomic oxygen would destroy the blanket material within a matter of days. The LEO prototype was therefore designed with a blanket of germanium-coated Kapton®, which would be resistant to atomic oxygen attack. This material is not as conducting as carbon but is still a weak conductor.

Three sample coupons were constructed that were as close to identical as possible except for the blanket material. One was made from uncoated Kapton®, and the other two had blankets coated with carbon and germanium, respectively. They were tested in a space simulation chamber for current collection as a function of applied bias voltage. As the results show, the highly insulating Kapton®-H, shown in Figure 3 by the curve designated "H," collected current linearly until around 300 volts. Current rose rapidly until about 400 volts when it became exponential, which is the signature of snapover. The weakly conducting germanium-coated blanket collected linearly only until about 125 volts when it began its rapid rise, and the much more conducting carbon blanket collected exponentially almost from the beginning. These experiments showed that the blanket itself could become involved in the snapover process and pointed to the critical need to test all proposed array coatings for plasma effects (Hillard, 1994). That is, with conductive blankets, the inherent conductivity can substitute for the secondary electron-induced conductivity to give snapover even at low voltages.

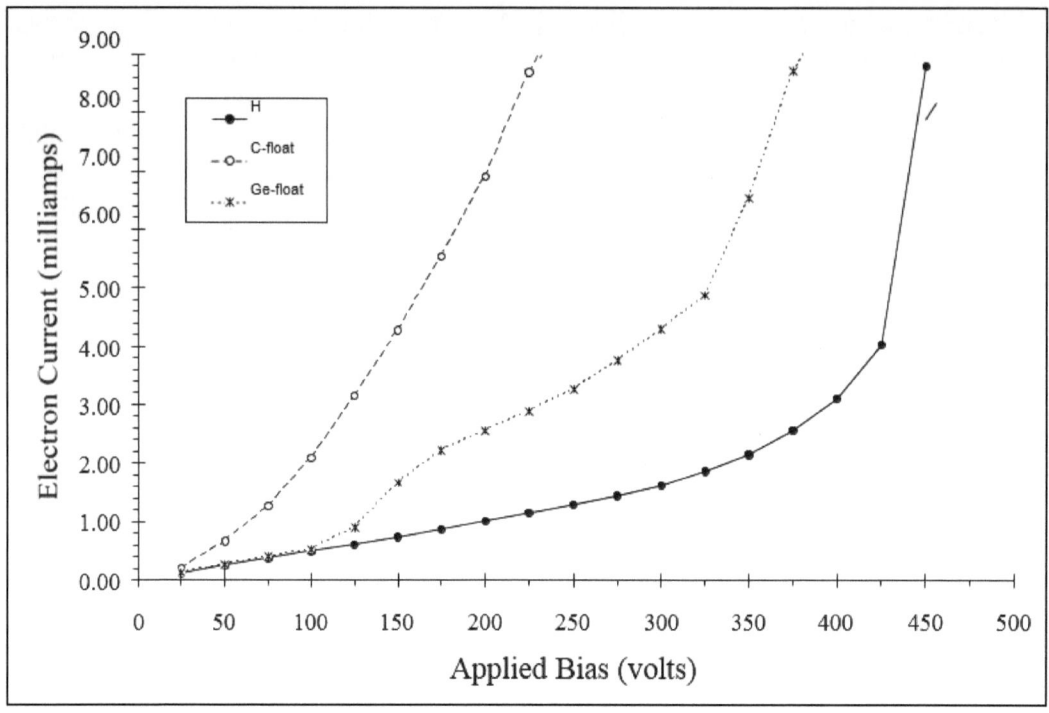

Figure 3—Electron Current vs. Bias for Three Solar Array Blanket Materials (Hillard, 1994)

5.2.1.3.3 Ion Collection

Ion collection for solar arrays is almost always a linear function of negative voltage. Again, the total array collection is the integrated value of all negative cells at their respective potentials relative to the ambient plasma, but for most solar arrays, this collection is small compared to ion collection from the structure. In the case of ISS, for example, Ferguson and Gardner (2002) could completely ignore solar array ion collection in modeling the ISS floating potential. When the array is in its wake, ion collection is further reduced.

5.2.1.3.4 Parasitic Power Drain

Biased components of a solar array exposed to the LEO plasma can collect current from the plasma. Positively biased exposed interconnects, for example, will collect electrons from the plasma. Such current collection is known as Parasitic Current Collection, or Leakage Current, as it represents a drain on the solar array power system. Khayms (2005) describes the plasma:

"… acting as a parasitic load on the power system in parallel with normal loads inside the spacecraft". In other words, power generated by the solar array is split between driving current through spacecraft loads and driving current through the plasma. In the case of a solar array with a large area of exposed conductors, the current collection (parasitic current) can consume a significant fraction of the solar array power and impact mission operations by substantially reducing the power available to the spacecraft systems.

Many efforts have taken place over the years to use the basic equations of plasma physics to estimate the magnitude of power system loss due to parasitic power drain, and one of them is presented here to illustrate the effect.

The high-voltage solar cell array for a high-power satellite looks more like a sheet electrode than like a spherical probe. K. L. Kennerud developed a method of analyzing the leakage current from such arrays based on fundamental equations developed by I. Langmuir (Kennerud, 1974). Kennerud's technique converts the linear array into a sphere having the same area, and then calculates the radius of the electron sheath surrounding the array. His experiments with small, positively charged solar-cell panels correlated well with his predictions.

Kennerud's results, shown in Table 2, Leakage Current from Positively Charged Solar Arrays, can be used to understand how the effect scales with altitude and electron density for the hypothetical solar array that he used.

Table 2—Leakage Current from Positively Charged Solar Arrays (Kennerud, 1974)

Altitude (km)	Electron Density (cm^{-3})	Electron Temperature (K)	Leakage Current		Power Loss, percent of Generated
			nA/cm^2	Amp per 1500V String*	
500	6×10^5	3,000	824.5	0.8494	7.72
700	2×10^5	3,000	274.8	0.2831	2.57
1,000	7×10^4	3,000	96.19	0.0990	0.90
2,000	2×10^4	3,200	28.38	0.0292	0.265
30,000	1×10^2	13,600	0.29	0.0003	0

* The string is 0.404 m by 255 m, with an area of 103.02 m^2.

Such rough calculations fail when the geometry becomes more complex. In particular, solar arrays with hidden interconnects such as the ISS arrays can collect current very differently from one with exposed interconnects. The ISS solar arrays, counter to intuition, collect more current at low electron temperatures than at high electron temperatures. Tests using an ISS solar array design at 300 to 500 V indicated that changes in the secondary electron yield properties of the dielectric materials affect the current collected at high electron densities and high bias voltages (Mikellides, 2005). Models have shown this phenomenon is caused by an electric field barrier to high-energy electrons (Mandell, et al., 2003). However, modeling electron collection by using spheres of equivalent "effective" area is very useful. Modern computer codes, such as the NASA Charging Analyzer Program (NASCAP) series, will provide useful estimates of parasitic power loss for any geometry. At high positive potential, snapover can make a solar array appear to be completely conductive. In addition, if a glow discharge caused by neutral gas ionization occurs on the array, the current collected can shoot up to tremendous levels (Ferguson, et al., 1998; Vayner, et al., 1999). Electric propulsion thrusters or plasma contactors, if placed in the vicinity of solar arrays, can short-circuit the plasma collection circuit and constitute a significant drain on the system power supply (Goebel, et al., 2014).

5.2.1.4 Current Collection at High Frequencies

NASA-HDBK-4007, Spacecraft High-Voltage Paschen and Corona Design Handbook, may be helpful for issues with multipaction and plasma effects involving high frequency power systems, where high frequency is defined as >50 kHz. For LEO plasma, the ion plasma frequency is 50 kHz. Ion response to changes above 50 kHz is limited. While significant new effects are not expected, most parameters of interest such as corona inception and extinction voltages are expected to exhibit frequency dependence. One effect did emerge in the early 1990s concerning insulated conductors energized with 20 kHz ac that were exposed to LEO plasma conditions (Button, et al., 1989). This work was underway because ISS was originally designed to use such a power system. Research was suspended when the ISS was reconfigured to use dc power.

If a conductor energized with low frequency ac is placed in LEO plasma, electrons are attracted to the insulating surface during the positive part of the cycle. These electrons "stick" to the material with a characteristic energy and are not repelled when the polarity changes to negative. Ions, however, are attracted during the negative part of the cycle and neutralize the electron charge for no net effect. At high frequencies, this neutralization process does not occur. Highly mobile electrons are still attracted during the positive part of the cycle; but ions, because of the much larger mass, cannot respond to the rapidly changing field. The outer surface therefore charges to a negative potential close to the peak voltage on the power system waveform and remains charged.

Although ions cannot respond to the rapidly changing voltage waveform, they do respond to the buildup of negative charge on the surface. The resulting ion flux results in equilibrium where the surface is charged, as a rule of thumb, to about 90 percent of the peak voltage level used in the system. For a high-voltage system, ions will easily acquire sufficient energy to sputter material from the insulation. Such charging can have a number of other implications that could include an arcing hazard, depending on where such surfaces are located with respect to other conductors.

5.2.1.5 Wake Effects

The motion of a spacecraft through its environmental plasma will produce a non-symmetric disturbance between the front and rear of the body. In LEO, the spacecraft velocity is much greater than the ion thermal speed but much less than the electron thermal speed. This flow regime is called mesothermal and also includes the case of supersonic flow, i.e., orbital speed greater than the ion acoustic speed. The plasma sheath on the upstream, or front, side of the spacecraft can be somewhat compressed compared to the stationary plasma case due to directed ion motion. However, the downstream side consists of various regions of plasma disturbances that extend over many body radii (Stone, 1981). A rarefaction wave initiates at the sheath edge and propagates away from the wake axis at an angle $\theta = \tan^{-1} (1/S)$, where S = Mach number = orbital speed/ion acoustic speed. Ions are swept out by the body, creating a near wake that is dominated by a much lower density region. The highly mobile electrons can enter the ion void region, but a negative space potential develops which acts to inhibit electron entry. An elevated electron temperature may be found in the near wake (Singh, et al., 1987). The near wake ends at about a Mach number of body radii downstream, but deviation from ambient density may extend much further. The morphology of the plasma wake region depends on several parameters: the ion acoustic Mach number, the spacecraft potential normalized by ambient electron temperature, the ratio of the characteristic spacecraft dimension to Debye length, and the electron to ion temperature ratio (Stone, 1981; Wright, 1988). The elongated wake region has a general symmetry about the wake axis, but complexity of spacecraft surface material and appendages can eliminate the symmetry near the spacecraft.

Instruments to measure the ambient plasma properties in LEO should be placed beyond the plasma sheath surrounding the structure and outside the wake of any structural element. Stone, et al. (1978) contains guidance for probe placement about a spacecraft. Samir, et al. (1986) contains a general discussion about the wake produced by spacecraft in LEO. See also Enloe, et al. (1997) that discusses the Space Shuttle experiment The Charging Hazards and Wake Studies (CHAWS).

6. ARCING

6.1 Solar Array Arcing

6.1.1 Background

For many years, the majority of spacecraft primary power systems used solar arrays and rechargeable batteries to supply 28 V. The choice of 28 V for the main bus voltage was made to take advantage of long-existing standards and practices within the aircraft industry. Plasma interactions at 28 V have not generally been considered a degradation factor of consequence. The only noted exceptions to their benign nature have occurred under extreme environmental conditions, especially during geomagnetic substorms for spacecraft operating at high inclinations. For low inclination spacecraft, i.e., those that completely avoid the auroral oval, 28-V systems have not been observed to arc.

As the power requirements for spacecraft increased, however, high-voltage solar arrays were baselined to minimize total mass and increase power production efficiency. With the advent of 100-V systems in the late 1980s, arcing began to be a significant concern (Ferguson, 1989). In 2011, after three decades of solar array arc testing, an international standard was created that defined a common set of terms and test methods associated with solar array arc testing. The ISO 11221 standard provides a common foundation for "Space solar panels – Spacecraft charging induced electrostatic discharge test methods" (ISO 11221, 2011). Therefore, this document will use the terminology and definitions described in the ISO 11221 standard. Where appropriate, parenthetical references to past terminology may be added to help the reader correlate items to older publications on the subject of solar array arcing.

Solar array arcs are generally characterized by the following parameters:

    a. Arc Inception Voltage (arc threshold, breakdown voltage) – The voltage required to initiate an arc depends on solar panel operating temperature, plasma current density, system bias voltage, insulation material properties, and construction and arrangement of the solar cells and solar cell strings. Arc inception voltage for a well-designed solar array can initiate as low as -75 V for spacecraft operating in a LEO plasma environment (Soldi, et al., 1997). Vayner, et al. (2001) have shown that discharge inception thresholds lower than about -300 V are invariably due to surface contamination with water and/or other contaminants.

    b. Temporal profile – The time from initiation to maximum current ranges from less than a microsecond to seconds, depending on the power source and the circuit impedance. The total duration of an arc ranges from microseconds to indefinite (seconds).

    c. Current profile – The arc current can be as large as 100 to 1,000 amperes depending on the capacitance of the solar array. See Figure 4, Peak Arc Current vs. Capacitance, from Purvis, et al., 1984.

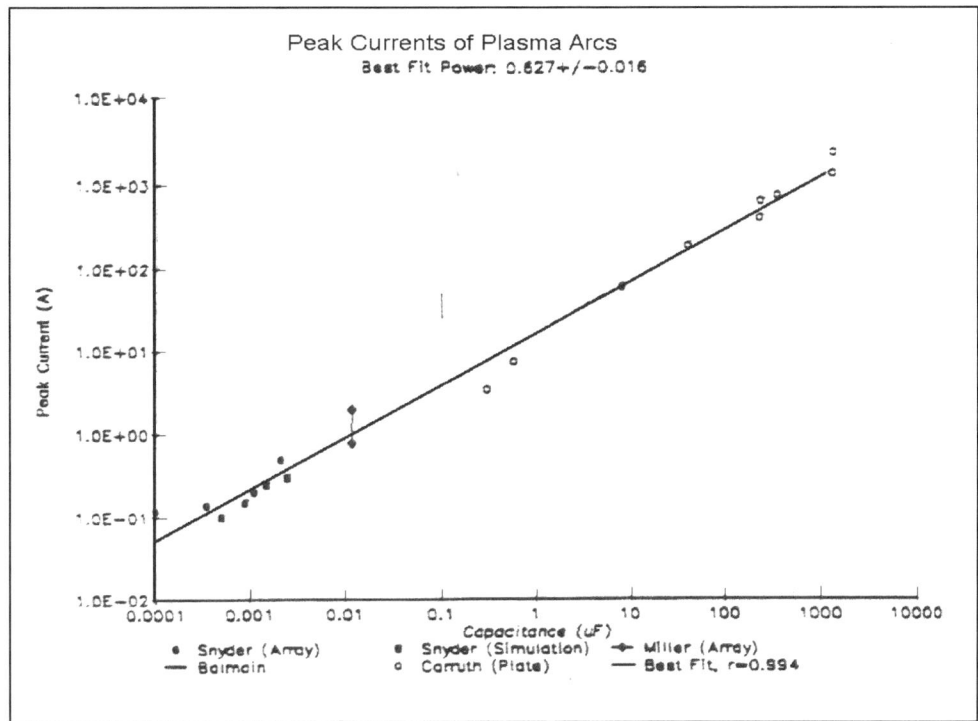

Figure 4—Peak Arc Current vs. Capacitance (Purvis, et al., 1984)

Based ISO 11221, there are two types of electrostatic discharges, i.e., arcs on solar arrays:

(1) Primary Arcs (fast transient or trigger arcs): The most common solar array arcs are characterized by rapid rise time followed by extinction in a time that is several times the rise time (e.g., 2 µs rise from zero to peak current with total duration of 10 µs). The critical parameter is that the energy involved is stored in whatever capacitance is available. The available capacitance can vary from a single array string to the entire spacecraft, depending on design. These arcs give rise to electromagnetic interference (EMI). Present understanding is that EMI does not cause permanent damage to materials, but it may damage or disrupt sensitive circuits. However, for spacecraft with multi-year operating lives, even for small arc rates, the number of arcs can add up into the 10,000's. Each low energy arc ablates a small amount of material, degrading the surface thermo-optical and electrical properties directly and from ablated material redepositing as contamination.

(2) Secondary arcs (sustained or continuous arcs): These are events that have been attributed with compromising or destroying on-orbit solar arrays. The process of forming a secondary arc begins with a primary arc injecting plasma in, or near, the interface between two solar array strings. The conductive plasma from the primary arc bridges the gap between strings, providing a low impedance path, and creating a short circuit for current to flow between the strings. The duration of

APPROVED FOR PUBLIC RELEASE—DISTRIBUTION IS UNLIMITED

secondary arcs is much longer than primary arcs; however, the peak current is limited to the maximum current through a string. In some cases, the duration of the secondary arc can be seconds to tens of seconds long, since the string power (not a finite capacitance) is directly feeding the arc. Based on a survey by Bodeau, secondary arcs with duration times greater than 1 ms should be classified as "sustained" arcs (Bodeau, 2014). In the worst-case scenario, a sustained arc can significantly damage the gap between two strings forming a permanent short circuit and consequently reducing the array's power output.

Primary and secondary arcs will be discussed in more detail in sections to follow. Since all arc events begin as a primary arc (fast transient), and most do not evolve beyond this phase, this type of arc has been the object of the most research in solar array arcing. Studies of the more destructive secondary arc have increased in number over the last decade as power levels have increased (causing higher and higher string voltages and currents to be used). The sections that follow are therefore organized around the primary (fast transient) discharge event. The secondary (sustained or continuous) arc will be addressed in the final section.

6.1.2 Arc Initiation Mechanism – Observations and Theories

The initiation of a solar array arc depends on the presence of a strong local electric field. Frequently, the source is an exposed interconnect which, depending on its location in the string, can be at high potential.

Most problematic are arcs that initiate at triple-points (conductor-dielectric-plasma interfaces). For a solar cell operating in LEO, this is usually the solar cell interconnect; but it can also be the edge of the solar cell (near the substrate or the coverslide). It has been shown that arcing on solar arrays at voltages less negative than about -1000 V is always mediated by the presence of a plasma. Identical samples to those that arced at -100 V in a plasma have been shown to withstand -1000 V bias in a pure (plasma-free) vacuum. Arcs that occur in a pure vacuum are called "vacuum arcs." Succeeding paragraphs discuss theories for the triple-point arcs that occur only in plasmas.

Arcs have been observed at relatively low potentials (as low negative as -75 V) when conductor surfaces are biased negative near insulator surfaces in the presence of a plasma (Soldi, et al., 1997). Arc rate is strongly dependent on plasma density and on coverslide temperature, which affects the surface conductivity. It can range from intermittent (on a scale of minutes and perhaps hours or longer) to several per second. Arc currents observed in ground tests are on the order of an ampere and can last several microseconds. These characteristics depend on the capacitance to space, increasing with increasing capacitance. These arcs are usually associated with solar cell interconnects, but have also been observed on biased conductor surfaces covered with dielectric strips. They are likely to be of concern whenever conducting surfaces at negative potentials with respect to plasma adjoin insulating surfaces.

Several mechanisms are proposed for initiation of the arcs. One mechanism proposes that a thin layer of relatively insulating film develops on the conductor. High electric fields develop across the film, caused by ion collection on the exposed face. The resulting electric field across the film causes electron emission from the conductor through the film into the plasma (Jongeward, et al., 1985). A second, though perhaps related, mechanism assumes that the high electric fields at the edge of the dielectric cause propagation of secondary electrons to the dielectric surface from near the conductor-dielectric-vacuum interface. Also, sufficiently intense electric fields can develop locally at the tips of structures built on the conductor surface because of the mobility of surface atoms driven by the electric field resulting from the presence of the nearby dielectric surface (Hastings, et al., 1992). Finally, gas desorbed from dielectric surfaces by electron impact can become ionized and serve as an ideal current path for the full-fledged arc.

At this time, no complete theories exist for the arc mechanism on solar cell arrays in a plasma. All require inclusion of an empirical factor to produce the observed low arcing voltage thresholds at triple-points. Experimental evidence indicates that an electron emission mechanism plays an important role in producing the arcs. A preliminary theory has been advanced that relates electron emission to the charging of a "dirty" layer on metal surfaces and the electric fields near an insulator-conductor-insulator surface configuration. This theory accounts for some of the experimental observations (Parks, et al., 1987).

An electron emission mechanism for solar array arcing is consistent with several experimental observations. Kennerud (1974) observed that the apparent ion collection of a solar cell array was enhanced by an order of magnitude prior to arcing. This could be accounted for either by electron emission, or by an increase in ion density of the plasma. Snyder and Tyree (1984) observed this emission as an increase in electron current collected by sensors in the vacuum chamber along with the solar array. They also noticed that these currents did not cease when the plasma generator was turned off. Arcing could still occur with no plasma in the tank as long as these emission currents were detected. Snyder (1984) also noticed that arcs did not take place in a very low-density plasma (10^2 cm^{-3}).

The occurrence of arcs can be predicted from the potential of the solar array coverslides relative to the plasma. In a very low-density plasma, even at relatively high bias voltages, the coverslides remained near plasma ground and no arcs occurred. At higher plasma densities, the coverslide potentials became several tens of volts more negative than plasma ground. When this condition existed, arcs occurred. Electrons from the plasma do not have enough energy to pass through the energy barrier set up by the biased interconnects and reach the insulator surfaces (Parks, et al., 1987). Electrons emitted from the interconnects of the array cause the coverslides to charge negatively relative to the plasma. These observations indicate that electron emission is necessary before the current pulse of the arcs can occur. Galofaro, et al. (1999) have shown that an arc is always preceded by a nanosecond burst of electrons from the arc site. This burst can also ignite arcs on nearby surfaces.

Jongeward, et al. (1985) proposed an arc mechanism model to account for this emission. The negatively biased interconnects tend to collect positive ions from the plasma. A layer of relatively high resistance material several angstroms thick can collect a sufficiently high surface density of positive ions to permit field emission of electrons from the region. This mechanism was first proposed to account for enhanced secondary electron yields from oxide films (Malter, 1936). Electrons emitted from this site are accelerated by the electric field between the cell, or interconnect, and the coverglass surface and strike the coverglass edge, which then emits secondary electrons in a cascade. Adsorbed gases are desorbed by electron impact. Ionization of these desorbed gases produces a dense plasma which is necessary for large currents to flow (Cho & Hastings, 1991). Some inferences can be made that are consistent with the experimental observations. There must be enough ion flux to the interconnect to maintain a high surface charge on the high resistance layer. The metal-insulator geometry provides a focusing effect which increases the ion flux to the interconnect and maintains the surface charge density. Field emission accounts for the relatively steady emission, which probably represents a metastable situation. The solar array arcs arise when this stability breaks down, producing increased electron emission.

This model predicts the time duration and current of the arcs to almost a factor of two. Progress is also being made in predicting arc rates using this model. For instance, Perez de la Cruz, et al. (1996) were successful in modeling the arc rates and thresholds seen in the SAMPIE experiment. The importance of adsorbed contaminants has been experimentally verified by Vayner, et al. (2002).

In 2004, Vayner, et al., summarized leading theories of arc initiation at conductor-dielectric-plasma interfaces as well as semiconductor-dielectric-plasma interfaces. Using experimental test data, Vayner, et al. (2004) added to existing models of arc initiation by noting the importance of energetic ions striking conductor (or semiconductor) surfaces and generating secondary electrons. The secondary electrons interact with the nearby dielectric surfaces giving rise to additional electron generation as well as desorption of gases resulting in the creation of a plasma. The plasma sheath formation leads to enhancement of the electric field at triple junction interface which can yield field emission electrons and drive substantial current flow in the plasma at the arc site.

By measuring the spectra of arc plasmas, Vayner, et al. (2008) showed that while vacuum arcs and plasma arcs do have some common features, there are noticeable differences that can be important in understanding plasma arc generation. Like plasma arcs at triple junction locations, vacuum arcs also show: cold emission of electrons at the cathode due to the high electric field, arc initiation on a small spot on the cathode surface, and traces of melted cathode material. However, due to differences in how the arcs are formed in plasma discharges, the plasma arc spectra show the influence of dielectric materials in the discharge formation as well as a population of high energy electrons not observed in vacuum arcs.

Brandhorst and Best (2001) have shown that solar array arcs can be initiated in the laboratory by simulated micrometeoroid strikes.

NASA-HDBK-4006A

6.1.3 Arc Inception Voltage (Arcing Threshold)

The arc inception voltage of a LEO solar array is difficult to predict as it is a function of a number of factors that include: coverglass thickness, interconnect type, cell geometry, cell operating temperature, room temperature vulcanized (RTV) insulation, substrate materials, plasma density, and many more.

Ferguson (1986) analyzed the arcing data from the Plasma Interactions Experiment II (PIX II) array and compared it to other ground and flight data (see Figure 5, Arc Rate vs. Voltage for Standard Interconnect Cells). Figure 5 is reproduced in Hastings (Hastings, et al., 1992; Hastings, 1995) with theoretical predictions superimposed. Ferguson noted that arc inception voltages (thresholds) could be established by examining arc rate data. A dramatic increase in arc rate signals the arc inception voltage (threshold) has been reached.

Studies by Upschulte, et al. (1994) and Hastings, et al. (1992) confirm that a voltage threshold exists for solar array arcing, and for certain values of a parameter called the field enhancement factor (FEF) (Cho, et al., 1990), reasonable values of the threshold are predicted. Vayner, et al. (2001) have shown that arcing is enhanced primarily by the presence of desorbing contaminant layers, although thin coverslides and other geometrical factors can also enhance the electric field and lower the arc inception voltage. Hot arrays (100 °C) have a higher arc inception voltage than cool arrays (room temperature) in ground tests, presumably because the coverslides become more conductive at high temperatures. These results were confirmed on orbit in the PASP Plus experiment for the APSA-type solar arrays (Soldi, et al., 1997).

In 2006, Hosoda, et al., added to the study of arc inception voltage through their development of a 400-V array for LEO. In addition to revisiting the mechanism for primary arc generation, they also showed the challenges of completely insulating a LEO solar array, including atomic oxygen interactions.

Likar, et al. (2007) showed the influence of solar cell shape, interconnect shape, and coverglass coatings on arc inception voltage. Included in the study was a full array module with 2,600 cells, which emphasizes the potential for variations due to workmanship in critical areas such as application of RTV insulation.

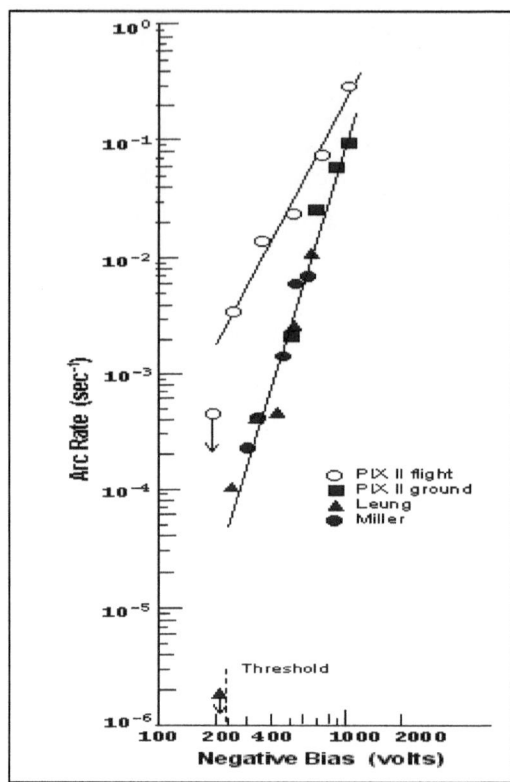

**Figure 5—Arc Rate vs. Voltage for Standard Interconnect Cells
(Threshold is inferred from the plasma arcing measurements. Ferguson, 1986)**

While many of the references in this section acknowledge -200 V as a common arc inception voltage, Vayner (2015) notes that low temperatures can significantly reduce arc inception voltages – down to the -100-V level.

As with parasitic current collection, the best way to determine arc inception voltage is to conduct laboratory tests on a representative sample of the array. Small changes in materials or layout of an array can have a pronounced impact on arc inception voltage when compared to a similar design.

6.1.4 Typical Waveform

Figure 6, Typical Waveform for an Array Arc, shows the time dependence of the current from an array segment during an arc (Snyder & Tyree, 1984). A typical arcing sequence has the following four regions:

    I. The arc is initiated and the current increases to a peak value. The rise time varies from less than 0.1 µs to about 1 µs. The peak amplitude and rise time depend primarily on the capacitance electrically connected to the arc site.

II. The current then remains near the peak value for some time.

III. The current decreases with a roughly exponential decay. The decay time associated with the termination of the arc should not be confused with the total duration of the arc. During this decay, the current is space-charge limited.

IV. The arc terminates suddenly and the array begins to recharge to the bias voltage. At this point, the coverslides of the array are substantially positive relative to both space and the arc point. The coverslides collect a substantial electron current from the plasma, resulting in the observation of a slight negative pulse.

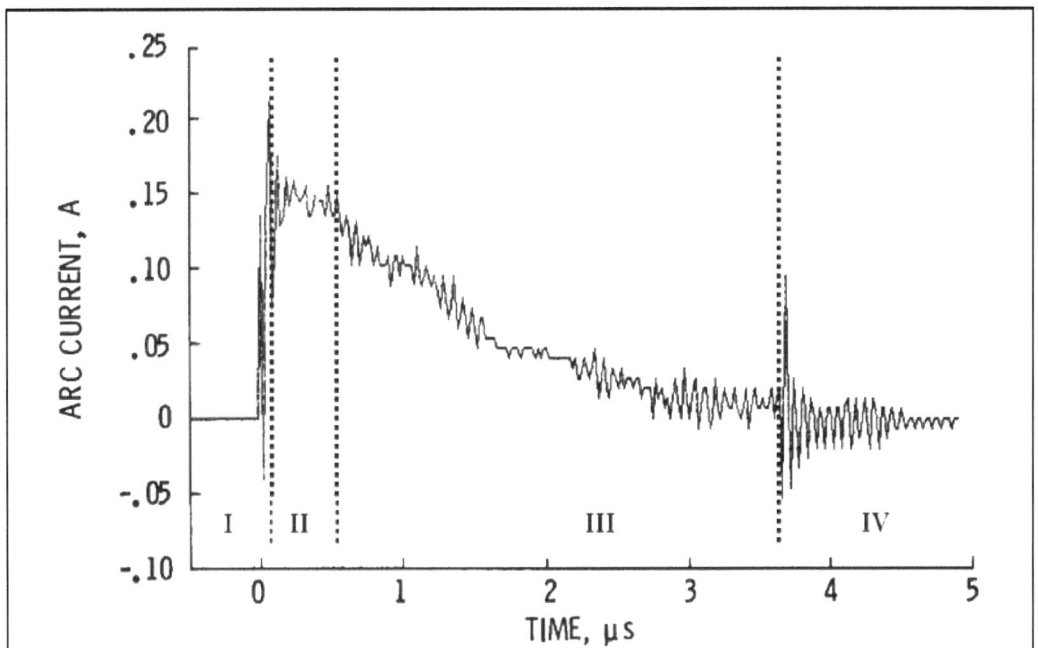

Figure 6—Typical Waveform for an Array Arc (Snyder & Tyree, 1984) Arc stages are: I- Onset, II – Peak Current, III-Decay, IV- Shut Off with ringing

6.1.5 System Response

Arc currents can flow into the surrounding plasma, with the return currents distributed over wide areas of other spacecraft surfaces.

During an arc, two things will happen. As charge leaves during an arc, the potential of the arc site changes and the potential of the system, electrically connected to the arc site, will change. As a result of the potential change, return currents will flow to restore equilibrium. The return currents will come both from the surrounding plasma and from the arc-generated plasma. There are two impacts on other systems. The structure currents will look like noise to instrumentation. And the change in spacecraft ground will affect plasma currents to surfaces. In principle, these responses are the same for transients of any cause: docking, thruster firings, waste dumps, and beam experiments. Only the magnitudes will be different.

The response of a system to an arc can be estimated from a circuit analysis, including terms to approximate the capacitances of the surfaces to space. An arc can be simulated in such a model by injecting an appropriate current pulse and computing the circuit transients (Metz, 1986).

6.1.6 Damage Due to Arcs

Initial indications that sustained arcs could cause substantial damage to solar arrays were obtained in testing where the bias power supply, intended to impress a potential difference between an array and its coverslides, was not sufficiently isolated from the sample when arcs occurred (see section 6.2.1). Tests at Lewis Research Center (LeRC), now Glenn Research Center (GRC), in the 1980s showed that solar array interconnects could be melted by arc currents as large as 40 A (Miller, 1985).

Although pictures of damage produced by on-orbit sustained arcs are rare because most arrays that have arced have not been recovered, we do have photos of damage suffered by the European Space Agency (ESA) European Retrievable Carrier (EURECA) spacecraft that was recovered by the Space Shuttle. Figure 7, Sample of Flight Array from ESA EURECA Mission after Sustained Arcing, shows a sustained arc site on its solar arrays. In this case, the sustained arc eventually burned through the array substrate to the grounded backing, completely shorting the array string to ground.

The Space Systems/Loral, LLC (SSL) satellites PAS-6 and Tempo-2 underwent sustained arcing in GEO that led to several shorted solar array strings and a severe loss of power. Although these were GEO failures, it is believed that after the initial arc occurs, the mechanism for sustained arcing is the same for LEO. Subsequent SSL satellites underwent extensive modification to prevent sustained arcing, and have had no similar string failures since that time. These modifications were the following:

    a. Changing the array layouts so that strings with high-voltage differences were not adjacent to each other.

    b. Including blocking diodes to prevent high currents from flowing during an arc.

    c. Grouting the cell edges on the strings with the highest voltage differences to prevent arcs from being sustained between strings.

A sustained arc on a test sample of arrays for the Earth Observing System – Morningside 1 (now Terra) (EOS-AM1) satellite, was seen in laboratory testing. Figure 8, Video Frame from EOS-AM1 Sustained Arc Test, is a frame from the videotape taken during the test; and Figure 9, Arc Site of Sustained Arc on EOS-AM1 Sample Array, shows the vicinity of the site where the arc occurred. The capacitor used in this test to start the initial arc was 5 microfarads, and the arc started and continued until the power supply was manually shut off seconds later. The solar array string was completely shorted out. This test led to rework of the entire array strings on the Terra satellite to prevent arcing on orbit. Flat-pack blocking diodes were incorporated into each string to prevent high currents from flowing during an arc, and Kapton® tape was used to cover exposed power bus conductors. The modifications made to the EOS-AM1 and SSL arrays are incorporated in NASA-STD-4005, section 4.1.5b.

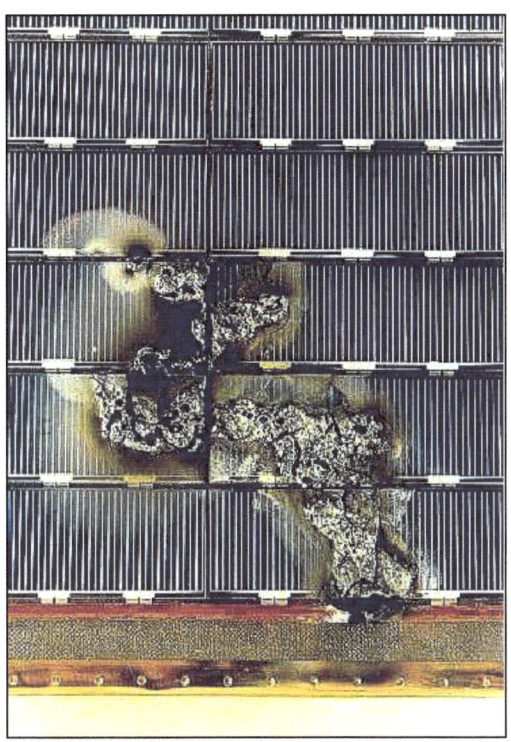

Figure 7—Sample of Flight Array from ESA EURECA Mission after Sustained Arcing (Ferguson and Hillard, 2003)

Figure 8—Video Frame from EOS-AM1 Sustained Arc Test
(Ferguson and Hillard, 2003)

Figure 9—Arc Site of Sustained Arc on EOS-AM1 Sample Array. Cells are 2x4 cm.
(Ferguson and Hillard, 2003)

6.2 Structure Arcing

Generally speaking, there are two forms of structure arcing. The first is triple-point arcing, as has been discussed for solar arrays; and the second is dielectric breakdown. For triple-point arcing, an insulator must surround a highly negative conductor; and an arc can occur at the conductor-insulator-plasma conjunction, where the electric field is highest. Dielectric breakdown is completely different, and will be discussed below.

An insulator not in the wake in LEO will achieve current balance at a potential within a few volts of the plasma potential. If that insulator covers a conductor, the conductor can be at a very different potential (such as the negative floating potential of the spacecraft, for example). In this case, a thin insulator can undergo dielectric breakdown under the high electric field developed across it. While this can occur for any type of insulator, it is of perhaps greatest interest in the case of anodized aluminum, the main ISS structural element, and a material used in the astronaut Extra-Vehicular Mobility Unit (EMU) (spacesuit). Because the dielectric layer in anodized aluminum is typically very thin (2.5-25 µm), it can break down at potentials as small as -100 V or less—less than the negative floating potential that is possible for a 160-V array. Figure 10, Anodized Aluminum Plate after Repeated Dielectric Breakdown Arcing, shows an aluminum witness plate anodized in the same batch as the ISS module exterior surfaces/hardware. The anodized witness plate has undergone repeated arcing in the laboratory with the ISS structure capacitance attached. Its thermal properties have been completely destroyed, along with most of the insulating surface layer of aluminum oxide (Carruth, et al., 2001). It was the arcing threat from the ISS anodized aluminum that forced ISS to incorporate the PCUs to control ISS floating potentials. The PCUs act by creating a large localized plasma cloud that makes good electrical contact with the surrounding plasma, and grounds the ISS structure to the ambient plasma. A generic plasma-contacting device is called a "plasma contactor."

Different samples of anodized material break down at different potentials in a plasma (Hillard, et al., 2000). ISS sulfuric acid anodized aluminum withstands about -200 V before breaking down. The ISS chromic acid anodized aluminum, with an oxide layer a tenth as thick as the sulfuric anodize, was found in ground tests to break down at about -72 V.

Dielectric breakdown currents will essentially discharge all surfaces close enough (about 2 meters or so) for the induced plasma cloud to reach. For thin dielectric layers, a few square meters of surface are effectively a capacitor of many microfarads, and can hold several joules of energy, all of which can be discharged in the arc. For many ISS surfaces, peak arc currents of hundreds of amps have been calculated. Arcs this strong will melt the arc site and spew molten metal through space. Plasma chamber tests of this kind of arcing are spectacular indeed! Arcs on one anodized surface have been seen to trigger arcs on nearby line-of-sight surfaces (sympathetic arcs; see Vayner, et al., 1998).

Figure 10—Anodized Aluminum Plate after Repeated Dielectric Breakdown Arcing (Schneider, et al., 2002b)

Very thin dielectric layers will have a low enough resistance that for the purposes of the plasma, they would collect current rather than building it up on their surfaces. Thus, while mitigating dielectric breakdown, they must be considered as conductors rather than insulators.

Predicting arc thresholds for thin insulating layers is not as simple as using the published dielectric strengths for insulating materials. It has been found that identical thicknesses of the same anodization can differ by a factor of three or more in arc threshold voltage in a plasma. This can be caused by differences in sealing the anodized surfaces, which could affect their resistance to plasma currents. Until the theoretical situation is better understood, plasma testing must be used to determine the dielectric strength of insulators in applications, which could lead to charging in LEO (Hillard, et al., 2000).

Carruth, et al. (2001) have found that dielectric breakdown can also be initiated by simulated micrometeoroid strikes at voltages as low as -75 V. In tests at the Glenn Research Center, anodized aluminum plates were seen to break down in a simulated space plasma at voltages as low as -55 V (Galofaro, et al., 1999).

6.2.1 Sustained Arcs

Arcs that occur in air when electrical contacts are made or broken are caused by breakdown of the neutral gas (Paschen discharge). Although these can become continuous, they are not the same phenomenon as the sustained (continuous) arcs in a LEO environment, which involve breakdown of the gas liberated by the arc itself. (See Holm (1999) for a discussion of continuous arcs in air.)

When the LEO arc circuit includes the solar arrays, power distribution cabling, or other source of power, it can be possible for structure or solar array arcs to become sustained (continuous). Such sustained arcs, fed by the power supply, i.e., solar array string power, have an essentially inexhaustible source of energy and can lead to catastrophic damage. This hypothesis for the loss of solar array strings on the SSL satellites PAS-6 and Tempo II was confirmed by ground tests done by Snyder, et al. (2000). Later testing on the EOS-AM1 arrays showed that sustained (continuous) solar array arcs could occur in a LEO environment at a string voltage as low as 100-120 V. (In those tests, the sustained arc occurred at a voltage relative to the surrounding plasma of -250 V.) The most recent data (Vayner, et al., 2003) has shown that strings with potentials as low as 40 V with respect to each other can lead to sustained arcing. Bodeau performed an extensive survey in 2014 which showed the voltage and current levels that give rise to sustained arcs [Bodeau, 2014]. The scenario for the catastrophic loss is given in Ferguson, et al. (1999), and is summarized here as follows:

First, a primary solar array arc must get started, usually at a triple-point as described above. In the case of the SSL arrays, the differential voltage between solar array and plasma could have been as low as 100 V, since the SSL arrays were using thin coverslides similar to the APSA cells, which arced at voltages as low as -75 V on orbit. See the PASP Plus results in Soldi, et al. (1997).

When the primary arc is generated, it discharges only the local capacitance, but the arc plasma expands out from the arc site and comes in contact with an exposed conductor at a very different voltage. In the case of the SSL arrays, the most positive end of the array strings was less than a millimeter away from the negative end. Now, the arc plasma makes direct contact with the other conductor and forms a very low resistance bridge to that spot. The arc current has changed from one that is discharging capacitance to a current between two ends of the solar array string. If the current available to the arc site from the functioning array is greater than a certain threshold value (believed to be about ½ amp for some array designs) and the voltage between strings is above a certain value (believed to be about 40 V for some array designs), the arc can become continuous. Bodeau (2014) has the most recent information on safe operating voltages and currents for solar arrays. Sustained arc thresholds are also a function of the separation between participating solar cells (Toyoda, 2008; and Boulanger, 2007). In ground tests, these arcs continued until the source of power was artificially turned off. In space, the arc would presumably continue until the exposed conductors were melted through and the circuit was thereby interrupted. This process could take seconds or minutes.

An arc that lasts long enough will locally heat the substrate and release gases. In the case of a Kapton® substrate, the Kapton® chars, but the char is also a good conductor, providing a path for the arc to continue. Snyder, et al. (2000) have shown that the heat generated in continuous arcs on Kapton® is sufficient to produce the Kapton® charring measured after the event.

In any event, a sustained arc can destroy a whole string (if the arc is between traces on the same string) or adjacent strings (if the arc is between strings) or the entire array power (if the arc is between combined power traces). The possibility of losing the entire array power on the Deep Space 1 mission caused the builders to remove a solar panel that had already been installed to modify it and its sister array to prevent continuous arcing. Its power traces were only a few millimeters apart, and were exposed both to the plasma and to each other before the modifications were made. Afterwards, insulating material was used to prevent arc plasma from shorting out between the power traces.

6.3 Tether System Arcs

An arc on an electrodynamic tether can become continuous. The arc on the TSS-1R tether that led to its break and the loss of the satellite was a continuous (sustained) arc with its power supplied by the tether. Figure 11, The End of the Remaining TSS-1R Tether, shows the burned, frayed, and broken tether end (Ferguson and Hillard, 2003). The arc site was a flaw in the tether insulation that evolved trapped gas, which became ionized and completed the arc circuit path (Szalai, et al., 1996; Vaughn, et al., 1997). In this case, the power source was constant voltage due to the 3500-volt potential difference at the switch in the tether deployer control boxes which caused the arc site to float at sufficient negative potential (about -600 V) necessary to keep the arc going and still collect ~1 A of electrons on the satellite.

TSS-1R had a mission goal of examining the power (or propulsion) potential of electrodynamic tethers. As such, high currents needed to be collected. During the planned flight objectives at maximum tether length, 0.5 A would be routinely collected; and a special flight operation to collect 0.75 A was planned before the tether was to be reeled in (Dobrowolny and Stone, 1994). Arc mitigation techniques as might be commonly used in flight, such as limiting the allowed current, could not be used. In the case of TSS-1R, mitigation of arc occurrence would have been to isolate the full tether electromagnetic force (emf) from the outgassing tether in the deployer system either by distance from switch or insulation. High voltage management requires thorough systems review and testing where possible.

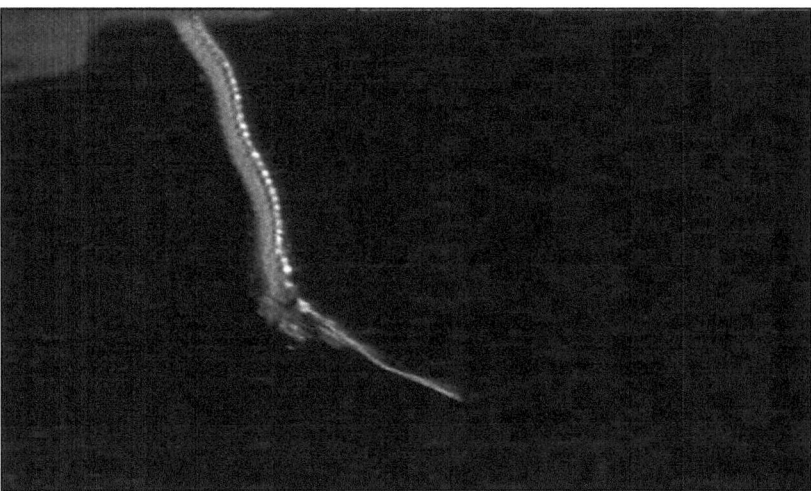

Figure 11—The End of the Remaining TSS-1R Tether (Ferguson and Hillard, 2003)

6.4 Arc-Generated Electromagnetic Interference (EMI)

Solar array arcs typically involve the discharge of very large currents for very short times. Not surprisingly, the electromagnetic spectrum associated with such discharges obeys the typical power law that has long been observed with arc discharges. An example of such a spectrum is shown in Figure 12, EMI from a Small Solar Array Arc and a Hypothetical ISS Anodized Aluminum Arc Compared to Orbiter's Specs (Leung, 1985). The test article was a small solar array sample that was proposed for a plasma interactions experiment in the Space Shuttle cargo bay. The test was designed to learn whether the radiated EMI from the sample would exceed orbiter specifications. The test was done with the bare array alone and with an added capacitance that simulated the energy storage associated with a full-size array. The biasing power supplies were electrically isolated from the arcs by a large resistor. As the curves show, even arcs from a small test array exceed allowed EMI specifications over most of the frequency range. It should be expected that arcing will always produce detectable EMI, and that laboratory testing will be needed to quantify the level of interference. The magnitude of radiated EMI is a strong function of the "antenna gain" composed of those conductive (radiating) elements connected to the arc site. This effect heavily influences the shape of the radiated EMI spectrum.

Since antenna gain is extremely difficult to estimate, testing is essential.

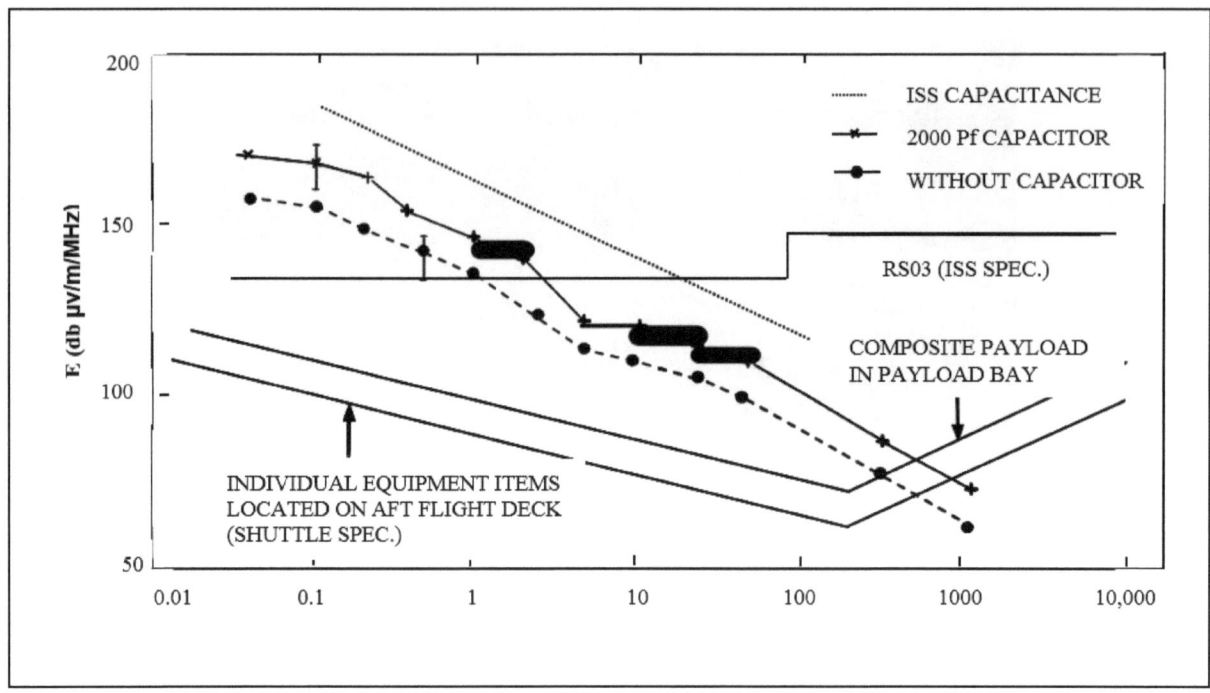

Figure 12—EMI from a Small Solar Array Arc and a Hypothetical ISS Anodized Aluminum Arc Compared to Orbiter's Specs (after Leung, 1985)

6.5 Risks of Arcing During EVA

Spacecraft charging can present a significant risk to an astronaut during EVA. Arcing can damage a space suit, and the resulting currents can flow through the astronaut ranging in levels from perception to fibrillation or greater. Contact with a surface that is at the spacecraft potential (either the structure floating potential or the solar array voltages) without arcing can still result in the spacesuit collecting currents from the ionosphere. A spacesuit that is attached to the spacecraft potential can itself become location for arcing, triggering other arcs locally. A spacesuit that is connected to the spacecraft can also become a source of charge collection with large bare metal surfaces and long conductive safety tethers.

Chromic acid anodized samples for astronaut EMUs were found to break down at potentials of only -60 V, relative to the plasma, with a two-sigma error bar of 10 V (Schneider, et al., 2002a). It is thus possible that an astronaut, grounded to ISS by his/her tether or conductive tools, could undergo an arc at only -50 V. A sneak circuit analysis showed that such arcs could put > 40 milliamps of current through an astronaut's heart (Koontz, 2005). Since this amount is enough to cause heart stoppage, it is imperative that, if the ISS plasma contactors are inoperable during astronaut extra-vehicular activities (EVAs) (spacewalks), a method be used to prevent ISS astronaut workplaces from floating more than 50 V negative.

For manned spacecraft, the possibility of an EVA must be part of the design trades. Charging controls and monitors need to be designed with the proper levels of redundancy as required for

APPROVED FOR PUBLIC RELEASE—DISTRIBUTION IS UNLIMITED

systems protecting against catastrophic hazards. For unmanned spacecraft where the possibility of EVA capture, service, and repair exists, the risks of spacecraft charging must be understood and documented so that proper operational controls can be implemented during service missions, particularly for those requiring EVA.

7. MITIGATION TECHNIQUES

7.1 Current Collection

If a spacecraft has no exposed high-voltage conductors, its current collection will be minimal. That is, insulation or encapsulation is a valid technique for preventing current collection. In LEO, the relatively high density space plasma will act to minimize the voltage variation across insulating surfaces and keep the potentials near the plasma potential, i.e., within a few kT_e of the plasma potential. If encapsulation or insulation is not possible, hiding conductive surfaces (like the edges of solar cells) from the ambient plasma by use of narrow spacing of overlying insulators (like coverslides) can choke off most current collection. Of course, if all high-voltage components are inside a sealed pressure vessel, they cannot collect current from the ambient plasma.

Encapsulation, or grouting with RTV rubber, of solar arrays has been shown to be an effective method to prevent electron collection and charging (Reed, et al., 2001). Of course, the grout must be UV and atomic oxygen (AO) resistant. Care must be taken in the use of encapsulants, however, when the possibility exists of outgassing in the presence of high-voltage components. For instance, on SAMPIE, one of the high-voltage power supplies was destroyed by a Paschen discharge that occurred on a high-voltage component where the encapsulant had delaminated and a neutral pressure was enclosed with the high-voltage component (Ferguson and Hillard, 1997). On TSS-1R, the "trigger arc" was a Paschen discharge due to entrained gas inside the tether pulley casings (Szalai, et al., 1996; Vaughn, et al., 1997).

Placing plasma-current-collecting conductors into the wake of a large spacecraft is an effective technique for preventing current collection. On ISS, for instance, data from the NASA/GRC Floating Potential Probe (FPP) instrument showed that when the arrays were turned into their own wakes, they collected such a small amount of electron current that the ISS structure would not charge. On ISS, this technique of wake-pointing the arrays is now used as a backup for the Plasma Contacting Units during astronaut EVAs. Of course, very high potentials on wake-pointing conductors can collapse the wake; but this will require thousands of volts potential for large structures. Wake pointing solar arrays for current collection control is rarely a viable approach due to the large loss in solar array energy generation through the LEO sun period.

7.2 Controlling Spacecraft Potential

There are four basic techniques to control spacecraft potential. The techniques (in no particular order) are:

1. Place the structure at the most positive potential generated by the LEO spacecraft power system (the positive ground option).

2. Ground the structure by brute force to the ambient plasma (the plasma contactor solution).

3. Prevent any plasma exposure of high-voltage conducting surfaces (the encapsulation solution).

4. Use electrical bonding as a mitigation strategy to prevent the spacecraft surfaces and structures from having different potentials with respect to each other and the plasma environment. Electrical bonding is addressed in MSFC-HDBK-3697, Electrical Bonding Design Guide Handbook, and NASA-STD-4003, Electrical Bonding for NASA Launch Vehicles, Spacecraft, Payloads, and Flight Equipment.

Mitigation strategies 1, 2, and 3 are discussed in order below. Short descriptions of variations on these methods are given in Ferguson, 2002.

7.2.1 Positive Ground

Since charging in LEO is dominated by electron collection on the most positive end of the solar arrays, and the negative end floats at about 90 percent (typically) of the string voltage, the positive end of the array will be about 10 percent of the array string voltage away from the plasma potential. For a 160-V array, this means a positively grounded structure will float at 16 V or less away from the plasma potential. Most deleterious plasma effects are minimal at such a potential. In fact, the structure in this case contributes to electron collection, and actually floats closer to plasma potential than the positive end of the array because of exposed grounded conductors on the structure.

However, most spacecraft power systems are negatively grounded because of a dearth of space-qualified electronics with the positive ground polarity. Although very efficient power management and distribution (PMAD) systems now exist that use buck-boost converters to change the ground polarity and voltage (Button, et al., 2002), most spacecraft busses do not incorporate this technology yet.

A variant of the positive ground technique uses a center-tapped array, but it will only cut the maximum structure potential to about half the solar array string voltage. Grounding the power system at about 90 percent of its maximum positive voltage would be nearly ideal, since it should place the spacecraft ground at near the plasma potential.

7.2.2 Plasma Contactors

A device that makes good contact with the surrounding plasma can effectively ground its point of contact. If the device is a large sheet of metal, it will dominate current collection and stay near plasma potential. However, the sheet of conductor must be much larger than the solar array effective electron-collecting area for this solution to work. The relative area of ion collection needed to balance the electron is simply the ratio of the electron thermal current density to the ram ion current density that is ~ 8. (See section 5.2.1.1 and equations 5 and 9.) In LEO, the drag produced by the addition of such a large area would be prohibitive.

Electron guns were used on PIX-II and PASP Plus (Purvis, 1985, and Guidice, et al., 1997) to emit the electrons being collected by high-voltage solar arrays and thus prevent charging, but such devices are limited by space charge considerations to low emitted electron currents. A better solution is a device that is not limited by space charge considerations, i.e., a plasma contactor.

A plasma contactor unit (PCU) generates a high-density plasma cloud, which expands and makes good electrical contact with the ambient plasma. Usually, a hollow cathode device emits a xenon plasma (Davis, et al., 1986) whose space charge is nullified by nearly equal densities of electrons and ions in the emitted cloud. The very mobile electrons carry current into the surrounding ambient plasma. This current can be very large. For instance, the PCU device on ISS can emit up to 10 amps of continuous electron current. In the case of ISS, the PCU can effectively ground the structure (to within about 40 V) to the ambient plasma. Away from the "grounding point" effects such as the motional emf will induce additional potentials which may only be an issue for very large structures such as the ~ 100 m long main truss on ISS.

While a hollow cathode plasma contactor requires gas vessels, plumbing, and refurbishment, other devices with little or no expellant are being explored for use as plasma contactors. As an example, a plasma contactor made of microtips and microscopic holes, with an imposed bias, could theoretically emit electrons over a wide area and thus defeat the space charge limitation with no working gas (or plasma). A patent has been awarded for using such a device to control spacecraft potentials in GEO (Katz, 2001), but making such a device work reliably in LEO is difficult because atomic oxygen interaction is detrimental to successful long-term operation.

7.2.3 Encapsulation

Encapsulating high-voltage conductors on solar arrays, etc., can have a two-fold beneficial effect. First, it can prevent arcing at triple-points by keeping the plasma away from the conductor-insulator junctions. Second, it can prevent electron collection by the arrays and thus prevent spacecraft charging at its root cause. Arrays that were ground tested in a simulated LEO plasma to withstand bias voltages greater than -300 V were those with the arrays or cells encapsulated (Reed, et al., 2001; Brandhorst and Best, 2001; Ferguson, et al., 2002). Since that time, other mitigation techniques have extended the arcing threshold to at least -500 V, but no arrays with unmitigated arcing have withstood more than -300 V to date. In 2012, a small satellite, fabricated by the Kyushu Institute of Technology (KIT) in Japan, with a 350-V array was launched for an on-orbit demonstration of high-voltage solar array technologies developed at KIT. In this array design, all exposed metallic parts were covered with S-691 silicon adhesive (Cho, et al., 2014).

NASA-HDBK-4006A

When encapsulating arrays or cells, one must not ignore several caveats. First, no air must be entrained anywhere. While this seems obvious, at least one set of encapsulated test arrays sent to NASA's GRC had sufficient air entrained that the coating delaminated and swelled under vacuum. In cases where only a very small amount of air is trapped, visible effects may not occur; yet the trapped air will present the danger of Paschen breakdown under high voltage.

Second, the encapsulant thickness must be sufficient to withstand dielectric breakdown at the highest array voltage. For thin-film arrays, this consideration can contribute significantly to the array mass. In keeping with the discussion on structure arcing, it is important that thin-film encapsulants be tested under voltage in a plasma environment, rather than relying solely on published dielectric strengths.

Third, the encapsulant must not peel away from high-voltage components, or Paschen breakdown can occur because of entrained outgassing products that can reach sufficiently high neutral pressures.

Fourth, the encapsulant must be able to withstand other aspects of the space environment for its design lifetime. Atomic oxygen, micrometeoroids and debris, UV, and X-ray exposure are some of the threats to the encapsulant. Glass stands up well to all of these environments while some plastics do not.

7.2.3.1 Vented Enclosures

It should be pointed out that the use of a sealed pressure vessel eliminates environmental interactions, and this applies to plasma interactions as well. In the more general case, high-voltage systems other than solar arrays are usually contained in a vented enclosure. To avoid plasma interactions, care must be taken that plasma does not enter the enclosure and react with exposed conductors inside. The key requirement on such systems is that all openings must be smaller than the plasma Debye length in the satellite orbit, which is ~ 0.5 cm at the F2 density peak. Openings in the experiment electronics enclosure must have smaller dimensions than this minimum to prohibit plasma interactions with the experiment electronics. Larger openings can be used if covered with an electrically connected conductive wire mesh of spacing less than the minimum Debye length. To provide a reasonable margin of safety, a general guideline is that no opening should exceed 0.10 cm in its largest dimension.

7.2.4 Arcing

7.2.4.1 On-Orbit Arc Detection

Usually, in ground tests of solar arrays under simulated LEO plasma conditions, and especially when the array can undergo sustained arcing, an arc detection circuit is employed. It essentially looks for a rapid positive change of the array or arc site potential toward the plasma potential, since this must happen when electrons are emitted during an arc. For example, a coil can be placed around the solar array string output wire, and changes in the coil current will indicate a transient in the line. Conversely, one can sense the emission of copious electrons and use this for arc detection. Further, the broadband EMI from an arc can be used for arc detection. In any event, electrical detection techniques can unambiguously detect arc occurrence. In ground tests, the power supply is electrically disconnected from the array to prevent the occurrence of sustained arcs that might damage or destroy the sample. Sometimes, the power supply is only disconnected when the arc continues for longer than 200 ms, for example, so that arcs that would be permanently sustained can be counted but are not allowed to cause damage on the sample. Such arc detection and array protection circuits can be built and used on solar arrays operating on orbit. If this is done, rather than totally preventing arcs, the damage to the arc site is limited or prevented. In this way, the arcs that do occur become tolerable.

The drawback to using an interrupting array protection circuit is the obvious disruption in spacecraft power each time an arc is detected. Rather than being the first line of defense against arcing, arc detection and array shunting must only be used when the disruptions they cause will be infrequent.

7.2.4.2 Prevention Techniques

The design of a solar array must consider the plasma environment and interactions with that environment. Arc prevention is extremely important. The following techniques have been shown in ground and flight tests to prevent arcs or minimize their damage:

a. If possible, use array string voltages of less than 55 V. No trigger arcs have been seen on LEO arrays of less than about 55 V string voltage or on anodized aluminum even under simulated micrometeoroid bombardment. Cold solar arrays coming out of eclipse will generate more voltage than when they operate at their maximum power point (at their hot equilibrium temperatures).

b. If solar array cell edges or interconnects are exposed to the LEO plasma and string voltages are greater than 55 V, the strings should be laid out on the substrate such that no two adjacent cells have a voltage difference of greater than 40 V. Sometimes, a leapfrog arrangement will be sufficient. In other high-voltage arrays, the strings should be arranged parallel to each other. Serpentine strings can be used to prevent the array width from becoming prohibitive. If the string layout cannot be modified to prevent cells with more than 40 V difference being adjacent to each other (anything less than about 1 cm can be considered adjacent), then the total string voltage must be kept low enough that the initial (trigger) arcs do not take place. The lowest known array trigger arcing has occurred on thin coverglass cells at about 75 V (PASP Plus results) (Soldi, et al., 1997).

c. For array string voltages greater than about 75 V, trigger arcs in LEO can be completely prevented by encapsulating the cell or array edges so they do not see the ambient plasma. The caveats mentioned above under "Encapsulation" in section 7.2.3 must be followed. If encapsulation is not possible, a thorough array bakeout on orbit (1 week at 100 °C or more) can get rid of contaminants and prevent trigger arcing up to about -300 V, or possibly more (see

Vayner, et al., 2002). Re-contamination can occur on "dirty" spacecraft (spacecraft with excessive venting, cold gas nozzles, etc.). Good encapsulation can prevent arcing up to 1000-V string voltage. However, environmental aging of the encapsulation material should be examined by ground test, which guides the best choice of materials.

  d. Sustained (or continuous) arcs can occur whenever trigger arcs occur and adjacent cells have more than 40 V potential differences. However, sustained arcs, in addition to this voltage threshold, have a current threshold below which they will not occur. See Bodeau (2014) for discussion of safe operating currents. If the current produced by each cell is above this threshold, a single string can sustain arcs. If each cell is below this current threshold, then isolating separate strings of solar cells from each other will prevent other strings from "feeding" the arc site and will prevent sustained arcs. This isolation can be achieved by using blocking diodes in each string. EOS-AM1, now called Terra, is an example (Snyder, et al., 2000). Care must be taken that the power bus and/or other components do not have the conditions necessary for sustained arcing. On the Terra arrays, for instance, it was found that diodes used to block inter-string currents did not prevent the bus power traces from having sustained arcing events. Covering all exposed bus conductors with Kapton® insulation finally solved the problem. Low-outgassing RTV can be used to cover bare conductors as well.

  e. RTV grout between adjacent solar cells and strings that have a high voltage with respect to each other has been shown to effectively block sustained arcs between cells and strings. The degree of coverage, etc., is important in determining the final voltage threshold for sustained arcing.

  f. Arrays of 300 V and greater string voltage must be fully encapsulated to prevent arcing. Caveats involved under "Encapsulation" in section 7.2.3 must be followed.

  g. Finally, although design and construction are important in preventing trigger arcs from evolving into sustained arcs, each new solar array design implementation must be verified by testing in a simulated LEO plasma chamber to ensure it will not arc. This is a critical step. The test bias voltage relative to the plasma should include the maximum array voltage when the arrays exit eclipse (or the highest floating potential expected on the spacecraft chassis). The inter-string voltage should be at least as great as that expected anywhere on the solar arrays on orbit. A test should be conducted at the low temperatures experienced at eclipse exit or from a sustained period of large sun off-pointing.

8. MODELING

8.1 Spacecraft Charging

The severity and widespread nature of plasma interactions have led to a considerable investment in the development of computer models. Many empirical and semi-empirical models are available with varying levels of capability and fidelity. Since the physics of current collection is fully embodied in Poisson's equation, a first-principles treatment is both possible and practical.

The standard such code that is available to U.S. citizens is NASCAP-2K (NASA Charging Analyzer Program), which is capable of modeling current collection and charging under LEO, GEO, interplanetary, and auroral conditions. NASCAP-2K was developed in conjunction with the U.S. Air Force as a follow-on to the original NASCAP computer program that dealt with spacecraft charging in geosynchronous orbit (Katz, et al., 1981; Mandell, et al., 1981; Rubin and Stevens, 1983).

NASCAP-2K incorporates lessons learned to-date, takes full advantage of modern computing power with much more sophisticated algorithms, and is designed for easier use. Capable of modeling current collection and charging under LEO, GEO, and auroral conditions, NASCAP-2K should now supersede both NASCAP and NASCAP/LEO (Neergaard, et al., 2001). NASCAP-2K is subject to International Traffic-in-Arms Regulations (ITAR) restrictions and at present cannot be given to non-U.S. citizens. For more information on distribution of these codes, see http://see.msfc.nasa.gov. European spacecraft charging modeling codes include the ESA Space Environment Information System (SPENVIS) family of codes, available on-line at http://www.spenvis.oma.be/spenvis/. The Japanese spacecraft charging analysis tool is called MUSCAT (Muranaka, et al., 2008).

To study ion collection by a high-voltage object in a spacecraft wake, the CHAWS (Charging Hazards and Wake Studies) experiment was flown aboard the WSF (Wake Shield Facility) on STS-60 (February 1994) and STS-69 (September 1995). Figure 13, NASCAP-2K Calculation of Plasma Potential and Density behind a Spacecraft at Orbital Speed, shows the *Object Toolkit* model of the WSF with CHAWS. The CHAWS probe is the off-center well-resolved rod seen toward the left of the figure, which shows the wake side of the WSF. Negative biases of up to 2 kV were applied to the probe, which was instrumented to measure both the total current and the distribution of current over the surface.

NASCAP-2K was used to compute the self-consistent space potentials and ion densities about WSF in a flowing low Earth orbit plasma, 1×10^{11} m^{-3} .1 eV, 7800 m s^{-1}. Ions are generated at the problem boundary with a thermal spread about the ram direction and are tracked until they strike the object or leave the simulation domain. Ion space charge is accumulated in the grid at each tracking step. The potential field of the probe extends nearly 0.4 m into the ram flow in the vicinity of the near edge of the WSF. The resulting electric fields cause the deflection of ions from this region into the wake where they may be collected by the probe, as seen in Figure 13. The grounded rear surface of the WSF, the WSF instrumentation, and the space charge contribution of ions deflected into the wake all play a role in screening the potential of the high-voltage CHAWS probe. As a result, the potential within the wake downstream of the probe falls off considerably faster than would be the case if the probe were in a cylindrical vacuum region. It is noteworthy that no ions strike the side of the probe nearest the edge of the WSF, but rather ions either strike the tip of the probe or else miss the tip and are attracted back to the side of the probe facing the center of the WSF.

Comparison of these calculations with the actual CHAWS measurements is discussed in Davis, et al., 1999.

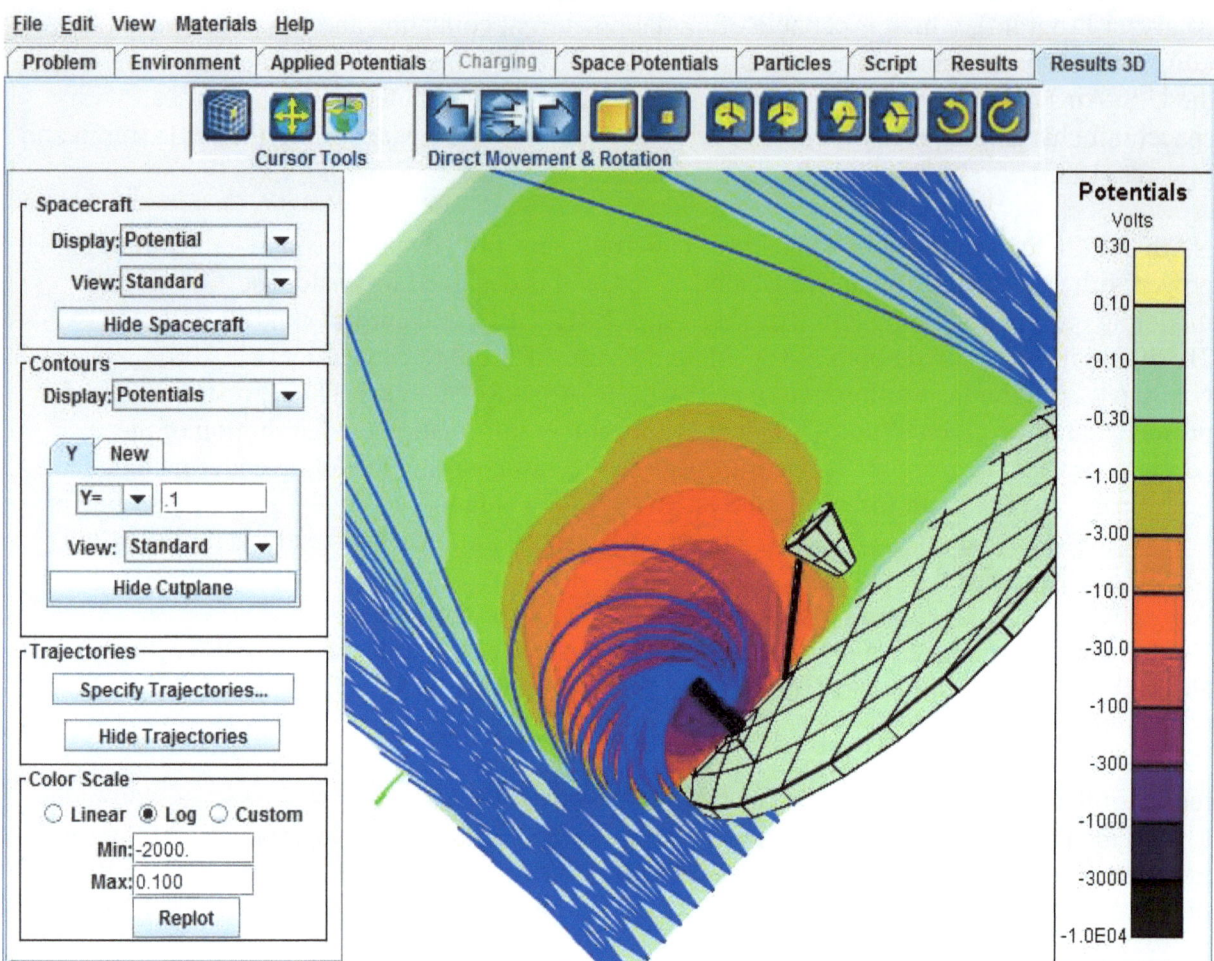

Figure 13—NASCAP-2K Calculation of Plasma Potential and Density behind a Spacecraft at Orbital Speed

8.1.1 Arcing

The process of electrical breakdown has not lent itself well to modeling, and electrical breakdown of solar arrays is no exception. The previously mentioned computer codes for determining potentials on all surfaces and electric fields in nearby space are certainly useful for solar arrays, but the actual initiation of an arc is extremely difficult to predict. Despite NASA's efforts to fund theoretical work in this area during the 1990s, no reliable model for arc initiation exists. Experience has shown that knowledge of the potential distribution is at best a rough indicator of the probability of an arc.

The complex geometries involved in cell construction and string layout along with the poorly understood properties of adhesives, coatings, and other materials often result in laboratory tests providing unexpected results. This emphasizes the need for testing of solar arrays in suitable space environmental chambers and ultimately as part of space experiments.

9. TESTING

The importance of testing in mitigating LEO spacecraft charging and its effects cannot be overstated (Ferguson, 1996). A valid LEO arc test must take place in a chamber whose pressure is less than about 33 milli-Pascal (250 micro-Torr) and in a plasma with an electron density of more than 10^5 electrons per cubic centimeter. The electron temperature should be less than about 2 eV but the lower the better, with on-orbit temperatures as low as 0.1 eV. To simulate the effect of ram ions, a streaming source with a few eV drifting ions may be employed. Rubin, et al. (2009) described a LEO plasma source capable of producing cold electrons and drifting ions. The sample temperature must be as low as the lowest sunlit temperature on orbit. To ensure that arcs will not occur in space, a sufficiently long waiting time must be used at each bias voltage that the arc rate is measured to be statistically significantly lower than the threshold arc rate. If the threshold is unknown, see Ferguson (1986) for a proper technique for establishing it in ground tests. Be aware that the arc rate at a given voltage usually decreases with time in the plasma; do not confuse this with an increasing arc voltage threshold (Ferguson, 1986). The chamber used for the tests should be big enough that the plasma sheath of the biased sample does not reach the chamber walls. For electron collection testing, the plasma source operating current must exceed the collected current level expected from the coupon under test. If this condition is not met, chamber plasma densities cannot be sustained and will drop below desired levels. Finally, use solar array design and building techniques that have been space qualified, whenever possible.

In LEO plasma testing, the array or anodized aluminum potential relative to the plasma (which in space is caused by spacecraft charging) is usually obtained by biasing the sample with a dc power supply. To investigate transient arcs, one must decouple the dc power supply from the arc current during an arc. This means the bias supply circuit must have a time constant greater than a few hundred microseconds, so the arc can build up and dissipate without being powered by the bias supply. This can be done by putting a large resistance in the arc circuit, and incorporating a capacitor to simulate the array or structure capacitance that would be discharged in the arc. For instance, if the on-orbit capacitance connected to the arc site is expected to be 0.1 microfarad, then this value capacitor can be used to provide current during the arc. With such a capacitor, the bias supply circuit can be given a 1-ms RC time constant (much greater than the arc time scale) with the use of a 10 kΩ series resistance. This effectively decouples the bias power supply from the arc. Of course, it also puts an upper limit on the arc rate attainable because of recharge time considerations.

In non-destructive sustained arc testing, the series resistance should be adjusted to limit the maximum current to that expected in the arc; and a cutoff circuit should be employed to shut off the bias supply after a few hundred microseconds. Experience shows that an arc that continues under such circumstances for more than about 200 μs will be sustained. Arc current and/or voltage waveforms should be closely monitored to distinguish between transient and sustained arcs. Videotapes of arc locations are helpful for diagnostic purposes. If destructive sustained arcs are allowed to occur, the videotape can confirm the arc time duration.

A comprehensive set of test procedures is documented in ISO 11221 (2011). This standard contains a set of internationally agreed upon terms to describe solar array arcs. The standard also provides suggested circuit arrangements from laboratories in the United States, Europe, and Japan.

NASA-HDBK-4006A

APPENDIX A

REFERENCES

A.1 Purpose and/or Scope

This Appendix provides reference information to the user. Ferguson and Hillard, 2003 served as the basis for the initial 2007 version of NASA-HDBK-4006. Ferguson and Hillard, 2003 contains an extensive annotated bibliography related to LEO spacecraft design. Additional resources to consult for spacecraft-plasma interactions are Tribble, 1995, and Hastings and Garrett, 1996.

A.2 Reference Documents

Bailey, G.J.; Balan, N.; Su, Y.Z. (1997). "The Sheffield University Plasmasphere Ionosphere model – a Review." *J. Atmospheric and Solar-Terrestrial Physics*, Vol. 59, Issue 13, pp. 1541-1552, Sept. 1997.

Bodeau, M. (2012). "Current and voltage thresholds for sustained arcs in power systems," *IEEE Transactions on Plasma Science*, 40, No. 2, pp. 192 – 200, 2012.

Bodeau, M. (2014). "Updated current and voltage thresholds for sustained arcs in power systems," *IEEE Transactions on Plasma Science*, 42, No. 7, pp. 1917 – 1921, 2014.

Boulanger, Bernard (2007). "Alcatel Alenia Space plasma ESD test results synthesis of solar array designs and solarbus flight heritage on satellite fleet". 10th Spacecraft Charging Technology Conference, Biarritz, France, June 18-21, 2007.

Brandhorst, H.; Best, S. (2001). "Hypervelocity Impact Studies on Solar Cell Modules." *Auburn University Report: AU-4-21839*, March 25, 2001.

BSR/AIAA G-003C-2010 (2010). American National Standard, "Guide to Reference and Standard Atmosphere Models," American Institute of Aeronautics and Astronautics.

Button, R.; Brush, A.; Sundberg, R. (1989). "Development and testing of a 20 kHz component test bed." IECEC-89. Proceedings of the Twenty-fourth Intersociety Energy Conversion Engineering Conference, Washington, D.C., August 6-11, 1989. New York: Institute of Electrical and Electronics Engineers. Vol. 1 (A90-38029 16-20), pp. 605- 610.

Button, R.M.; Kascak, P.E.; Lebron-Velilla, R. (2002). "Digital Control Technologies for Modular DC-DC Converters." NASA/TM-2002-211369, Feb. 2002.

Carruth, Jr., M.R.; Vaughn, J.; Holt, J.M.; Werp, R.; Sudduth. R.D. (1992). "Plasma Effects on the Passive External Thermal Control Coating of Space Station Freedom." AIAA 92-1685, AIAA Space Programs and Technologies Conference and Exhibit, Huntsville, AL, March 24-27, 1992.

Carruth, Jr., M. R.; Schneider, T.; McCollum, M.; Finckenor, M.; Suggs, R.; Ferguson, D.; Katz, I.; Mikatarian, R; Alred, J.; Pankop, C. (2001) "ISS And Space Environment Interactions Without Operating Plasma Contactor." AIAA-2001-0401, 39th Aerospace Sciences Meeting and Exhibit, Reno, NV. Jan 8-11, 2001.

Chappell, C.R. (1972). "Measurements of the Morphology and Dynamics of the Plasmasphere," *Review of Geophysics and Space Physics*, Vol. 10, No 6, pp. 951-979.

Chen, F.F. (1965). "Electric Probes," in *Plasma Diagnostic Techniques*, R.H. Huddlestone and S.L. Leonard, Eds., Academic Press, New York, NY, pp. 113-200.

Chen, F. F. (1984). *Introduction to Plasma Physics and Controlled Fusion*, Plenum Press, New York 1984.

Cho, M.; Hastings, D.E.; Kuninaka, H. (1990). "Dielectric charging process in high voltage solar cell arcing." 17th International Symposium on Space Technology and Science, Tokyo, Japan, May 20-25, 1990. Vol. 2 (A92-5345123-12) pp. 1421-1426.

Cho, Mengu; Hastings, D.E. (1991). "Dielectric charging processes and arcing rates of high voltage solar array", AIAA-91-0605, 29th Aerospace Sciences Meeting, Reno, NV, Jan 7-10, 1991.

Cho M.; Masui, H.; Iwai, S.; Yoke, T.; Toyoda, K. (2014). "Three hundred fifty volt photovoltaic power generation in low earth orbit," *Journal of Spacecraft and Rockets*, Vol. 51, No. 1, pp. 379-381, Jan.-Feb. 2014.

"Computational procedure used in the development of the MSFC modified Jacchia model atmosphere." (1970). CE Environment criteria guidelines for use in space vehicle development, MSFC. SEE N70-40876 23-30.

Craven, P. D.; Wright, Jr., K. H.; Minow, J. I.; Coffey, V. N.; Schneider, T. A.; Vaughn, J. A.; Ferguson, D. C.; Parker, L. N. (2009). "Survey of International Space Station Charging Events," AIAA-2009-0119, 47th AIAA Aerospace Sciences Meeting and Exhibit, Orlando, FL, Jan. 5-8, 2009.

Davis, V.A.; Katz, I.; Mandell, M.J.; Parks, D.E. (1986). "Three dimensional simulation of the operation of a hollow cathode electron emitter on the Shuttle orbiter." International Conference on Tethers in Space, Arlington, VA, Sept. 17-19, 1986.

Davis, V.A.; Gardner, B.M.; Guidice, D.A. (1998). "PASP Plus Solar Array Parasitic Current Collection Flight Results." *IEEE Transactions on Plasma Science*, Vol. 26, No. 1, February 1998, pp. 46-58.

Davis, V. A.; Mandell, M. J.; Cooke, D. L.; Enloe, L. (1999). "High-Voltage Interactions in Plasma Wakes: Simulation and Flight Measurements from the Charge Hazards and Wake Studies (CHAWS) Experiment," *Journal of Geophysical Research 104*, 12445-12459, 1999

de La Beaujardière, O. (2004). "C/NOFS: A mission to forecast scintillations," *Journal of Atmospheric and Solar-Terrestrial Physics*, Vol 66 (17), pp. 1573-1591.

Dobrowolny, M.; Stone, N. H. (1994). "A technical overview of TSS-1: the first tethered satellite system," *Il Nuovo Cimento*, 17C, p. 1, 1994.

Dunbar, W.G. (1988). "Design Guide: Designing And Building High Voltage Power Supplies." Air Force Wright Aeronautical Laboratories: AFWAL-TR-88-4143, Vol. 2.

Enloe, C. L.; Cooke, D. L.; Pakula, W. A.; Violet, M. D.; Hardy, D. A.; Chaplin, C. B.; Kirkwood, R. K.; Tautz, M. F.; Bonito, N.; Roth, C.; Courtney, G.; Davis, V. A.; Mandell, M. J.; Hastings, D. E.; Shaw, G. B.; Giffin, G.; Daga, R. M. (1997). "High-voltage interactions in plasma wakes: Results from the charging hazards and wake studies (CHAWS) flight experiments," *J. Geophys. Res.*, 102, A1, pp. 425-433, 1997, DOI 10.1029/96JA02931.

Ferguson, D.C. (1986). "The voltage threshold for arcing for solar cells in LEO: Flight and ground test results." NASA TM 87259, 1986.

Ferguson, D.C. (1989). "Solar Array Arcing in Plasmas", NASA CP-3059, Third Annual Workshop on Space Operations Automation and Robotics (SOAR 1989), Houston, TX pp. 509-513, July 25-27, 1989.

Ferguson, Dale C. (1991) "LEO Space Plasma Interactions", NASA CP-3121, Space Photovoltaic Research and Technology 1991, Cleveland, Ohio, May 7-9, 1991, section 47.

Ferguson, D.C. (1996). "The Role of Space Plasma Simulation Chambers in Spacecraft Design and Testing." Thirty-first Intersociety Energy Conversion Engineering Conference Proceedings, Institute of Electrical and Electronics Engineers, pp. 2188-2192.

Ferguson, D.C. (2002). "Alternatives to the ISS Plasma Contacting Units," NASA/TM-2002-211488, May 2002.

Ferguson, Dale C.; Gardner, Barbara (2002). "Modeling International Space Station (ISS) Floating Potentials," NASA TM-2002-211487, May 2002.

Ferguson, D.C.; Hillard, G.B. (1997). "Lessons for Space Power System Design from the SAMPIE Flight Experiment." AIAA Paper #97-0087, 35th Aerospace Sciences Meeting and Exhibit, Reno, NV, January 6-10, 1997.

Ferguson, D.C.; Hillard, G.B. (2003). "New NASA SEE LEO Spacecraft Charging Design Guidelines—How to Survive in LEO Rather Than GEO." NASA/TP—2003-212737. Dec. 2003.

Ferguson, D.C.; Hillard, G.B.; Snyder, D.B.; Grier, N.T. (1998). "The Inception of Snapover on Solar Arrays: A Visualization Technique." AIAA Paper #98-1045, 36th Aerospace Sciences Meeting and Exhibit, Reno, NV, Jan 12-15, 1998.

Ferguson, D. C.; Hillard, G. B.; Vayner, B. V.; Galofaro, J. T. (2002). "High Voltage Space Solar Arrays." 53rd International Astronautical Congress of the International Astronautical Federation (IAF), Houston, TX, Oct. 10-19, 2002, IAC Paper 02-IAA.6.3.03.

Ferguson, Dale C., Snyder, David B., and Carruth, Ralph (1990). "Findings of the Joint Workshop on Evaluation of Impacts of Space Station Freedom Grounding Configurations", Space Environment Analysis Workshop, Noordwijk, Netherlands, October 9-12, 1990, NASA Technical Memorandum 103717.

Ferguson, D.C.; Snyder, D.B.; Vayner, B.V.; Galofaro J.T. (1999). "Array arcing in orbit – From LEO to GEO." AIAA Paper #99-0218 presented at 37th Aerospace Sciences Meeting and Exhibit, Reno, NV, Jan. 11-14, 1999.

Ferguson, D. C.; Vayner, B. V.; Galofaro, J.T.; Hillard, G. B.; Vaughn, J.; Schneider, T. (2005). "NASA GRC and MSFC Space-Plasma Arc Testing Procedures." 9th Spacecraft Charging Technology Conference, Tsukuba, Japan, April 4-8, 2005.

Gabdullin, F. F.; Korsun, A. G.; Lavenko, E. G.; Mitroshin, A. S.; Tverdokhlebova, E. M. (2007). "The plasma plume of the ISS plasma contactor unit under the effect of the geomagnetic field," 30th International Electric Propulsion Conference, Florence, Italy, 17-20 Sep, 2007, IEPC-2007-049.

Gallagher, D.L.; Craven, P.D.; Comfort, R.H. (2000). "Global Core Plasma Model," *Journal of Geophysical Research*, Vol. 105, pp. 18819-18833.

Galofaro, J.T.; Doreswamy, C.V.; Vayner, B.V.; Snyder, D.B.; Ferguson, D.C. (1999). "Electrical Breakdown of Anodized Structures in a Low Earth Orbital Environmental," NASA/TM-1999-209044, July 1999.

Garrett, H.B.; Mullen, E.G.; Ziemba, E.; DeForest, S.E. (1978). "Modeling of the Geosynchronous Plasma Environment, 2 ATS-5 and ATS-6 Statistical Atlas." Rep. AFGRL-TR-78-0304. Air Force Geophys. Lab.

Gilchrist, B. E.; Bonifazi, C.; Bilen, S. G.; Raitt, W. J.; Burke, W. J.; Stone, N. H.; Lebreton, J. P. (1998). "Enhanced electrodynamic tether currents due to electron emission from a neutral gas discharge: Results from TSS-1R mission," *Geophysical Research Letters*, 25, p. 437, 1998.

Goebel, Dan M.; Filimonova, Olya; Anderson, John R.; Katz, Ira; Leifer, Stephanie; Polk, James E. (2014) "Definitive High Voltage Solar Array Testing in Space and Thruster Plume Plasma Environments," 13th Spacecraft Charging Technology Conference, Pasadena, CA, 23–27 June 2014.

Grier, N.T.; Stevens, N.J., (1978). "Plasma Interaction Experiment (PIX) Flight Results," Proceedings Spacecraft Charging Technology – 1978, pp. 295-314, AFGL-TR-79-0082 and NASA CP-2071, 1979.

Grier, N.T. (1985). "Plasma interaction experiment II : laboratory and flight results," Spacecraft Environmental Interaction Technology -1983, NASA CP-2359, pp. 333-347, 1985

Grossi, M. D. (1995). "Plasma Motor Generator (PMG) electrodynamic tether experiment," NASA-CR-199523, June 1995.

Guidice, D.A.; Davis, V.A.; Curtis, H.B.; Ferguson, D.C.; Hastings, D.E. (1997). "Photovoltaic Array Space Power Plus Diagnostics (PASP Plus) Experiment," Massachusetts Inst. of Tech. report. AD-A331959 PL-TR-97-1013, Mar 01, 1997.

Hastings, D.E. (1995). "A review of plasma interactions with spacecraft in low Earth orbit," Journal of Geophysical Research. Vol. 100, No. A8, pp. 14457-14483.

Hastings, D. E.; Cho, M.; Kuninaka, H. (1992). "The arcing rate for a High Voltage Solar Array – Theory, experiment and predictions," *Journal of Spacecraft and Rockets*. Vol. 29, Issue 4, pp. 538-554, July-August 1992.

Hastings, D. E.; Garrett, H. (1996). *Spacecraft-environment interactions*, Cambridge University Press, ISBN 978-0521607568, 1996.

Hedin, A. (1987). "MSIS-86 thermospheric model." *Journal of Geophysical Research*. Vol. 92, pp. 4649-4660.

Hickey, M. (1988). The NASA Marshall engineering thermosphere model. NASA-CR-179359, July 1988.

Hillard, G.B. (1994). "Plasma chamber testing of advanced photovoltaic solar array coupons." *Journal of Spacecraft and Rockets*, Vol. 31, Issue 3, pp. 530-532.

Hillard, G.B.; Bailey, S.G.; Ferguson, D.C. (2000). "Anodized Aluminum as Used for Exterior Spacecraft Dielectrics." 6th Spacecraft Charging Technology Conference, Hanscom Air Force Base, MA, 2000. AFRL-VS-TR-20001578, pp. 111-113.

Holm, R. (1999). *Electric Contacts: Theory and Applications*. Fourth edition, Springer Verlag. 1999.

Hosoda, S.; Okumura, T.; Kim, J. H.; Toyoda, K.; Cho, M. (2006). "Development of 400 V Solar Array Technology for Low Earth Orbit Plasma Environment," in IEEE Transactions on Plasma Science, vol. 34, no. 5, pp. 1986-1996, Oct. 2006.

ISO 11221 (2011), Space systems – Space solar panels – Spacecraft charging induced electrostatic discharge test methods, 2011. Reference number ISO 11221:2011(E).

ISO TR 11225 (2012), Space Environment (Natural and Artificial) – Guide to Reference and Standard Atmosphere Models. Reference number ISO/TR 11225:2012.

Istomin, V. G. (1966). "Observational results on atmospheric ions in the region of the outer ionosphere," *Annals of Geophysics,* 22, 255.

Iwasa, M.; Tanaka, K.; Sasaki, S.; Odawara, O. (2006). "Study of the Plasma Interference With High-Voltage Electrode Array for Space Power Application," *IEEE Transactions on Plasma Science*, Vol. 34, No. 5, pp. 1997-2003, Oct. 2006.

Johnson, C.Y. (1969). "Ion and Neutral Composition of the Ionosphere," Annals of the IQSY, Vol. 5, Solar-Terrestrial Physics: Terrestrial Aspects, pp. 197-214. Massachusetts Institute of Technology, published by the MIT Press.

Jongeward, G.A.; Katz, I.; Mandell, M.J.; Parks, D.E. (1985). "The role of unneutralized surface ions in negative potential arcing." *IEEE Transactions on Nuclear Science*, Vol. NS-32, pp. 4087-409.

Katz, Ira (2001). "Spacecraft solar array charging control device," U.S. Patent #6,177,629.

Katz, I.; Cassidy, J.J.; Mandell, M.J.; Parks, D.E.; Schnuelle, G.W.; Stannard, P.R.; Steen, P.G. (1981). "Additional application of the NASCAP code. Vol. 1: NASCAP extension." NASA-CR-165349, February 1981.

Kennerud, K.L. (1974). "High voltage solar array experiments," NASA-CR-121280, March 1974.

Khayms, Vadim; Logan-Garisch, Anthony; Kannenberg, Keith (2005). "Measurements and Modeling of a Solar Array Floating Potential and Leakage Current in a Hall Thruster Plume Environment", AIAA 2005-3862, 41st AIAA/ASME/SAE/ASEE Joint Propulsion Conference, Tucson, AZ, 10-13 July 2005.

King, R.L. (1978). "A computer version of the US Standard Atmosphere," NASA-CR-150778, August 1978.

Koontz, S. (2005). "EVA Shock Hazard (ISS-EVA-312) Assessment and Control: 1) Strategy, Methods, and Forward Plan. 2) VIPER/PHALCON Role and Requirements," ISS VIPER Working Group.

Leslie, F.W.; C.G. Justus (2011). "The NASA Marshall Space Flight Center Earth Global Reference Atmospheric Model – 2011 Version," NASA-TM-2011-216467, June 2011.

Leung, P. (1985). "Characterization of EMI generated by the discharge of a VOLT solar array," NASA-CR-176537, November 1985.

Likar, J.J.; Bogorad, A.L.; Vayner, B.V.; Galofaro, J.T. (2007). "Influence of Solar Cell Shape, Interconnect Shape, and Coverglass Coatings on Solar Array Arcing Parameters." 2007 IEEE Radiation Effects Data Workshop, Honolulu, HI, 2007, pp. 26-29.

Malter, L. (1936). "Anomalous Secondary Electron Emission A New Phenomenon," *Phys. Rev.*, Vol. 49, 478.

Mandell, M.J.; Katz, I.; Stannard, P.R. (1981). "Additional extensions to the NASCAP computer code, Vol. 1," NASA-CR-167855, October 1981.

Mandell, M. J.; Davis, V. A.; Gardner, B. M.; Jongeward, G. A. (2003). "Electron collection by International Space Station solar arrays," 8th Spacecraft Charging Technology Conf., Huntsville, AL, 20-24 October 2003, NASA/CP-2004-213091.

Metz, R.W. (1986). "Circuit Transients Due to Arcs on a High Voltage Solar Array," *Journal of Spacecraft and Rockets*, Vol. 23, Issue 5, pp. 499-504.

Mikellides, Ioannis G., Jongeward, Gary A., Schneider, T., Peterson, T., Kerslake, T.W., and Snyder, D. (2005) "Solar Arrays for Direct-Drive Electric Propulsion: Electron Collection at High Voltages". *Journal of Spacecraft and Rockets,* Vol. 42, No. 3, May–June 2005.

Miller, W.L. (1985). "An investigation of arc discharging on negatively biased dielectric conductor samples in a plasma." *Spacecraft Environmental Interactions Technology*, pp.367-377 (SEE N85-22470 13-18).

Minow, J. I.; Wright, Jr., K. H.; Chandler, M. O.; Coffey, V. N.; Craven, P. D.; Schneider, T. A.; Parker, L. N.; Ferguson, D. C.; Koontz, S. L.; Alred, J. W. (2010). "Summary of 2006 to 2010 FPMU Measurements of International Space Station Frame Potential Variations," 11th Spacecraft Charging and Technology Conference, Albuquerque, NM, 20-24 September, 2010.

MSFC-HDBK-3697 (2014), Electrical Bonding Design Guide Handbook, 2014.

Muranaka T.; Hosoda, S.; Kim, J.H.; Hatta, S.; Ikeda, K.; Hamanaga, T.; Cho, M.; Usui, H.; Ueda, H. O.; Koga, K.; Goka, T. (2008). "Development of Multi-Utility Spacecraft Charging Analysis Tool (MUSCAT)," *IEEE Transactions on Plasma Science*, 36, No. 5, p. 2336 – 2349, 2008.

NASA-HDBK-4002A (2011). Mitigating In-space Charging Effects – A Guideline.

NASA-HDBK-4007 (2016) Spacecraft High-Voltage Paschen and Corona Design Handbook.

NASA-STD-4003A (2013), Electrical Bonding for NASA Launch Vehicles, Spacecraft, Payloads, and Flight Equipment.

NASA-STD-4005A (2016), Low Earth Orbit Spacecraft Charging Design Standard.

Neergaard, L.E.; Minow, J.; McCollum, M.; Cooke, D.; Katz, I.; Mandell, M.; Davis, V.; Hilton, J. (2001). "Comparison of the NASCAP/GEO, POLAR, SEE Charging Handbook, and NASCAP-2K.1 Spacecraft Charging Codes." 7th Spacecraft Charging Technology Conference Noordwijk, April 23-27, 2001.

Parks, D.E.; Jongeward, G.A.; Katz, I.; Davis, V.A. (1987). "Threshold-determining mechanisms for discharges in high-voltage solar arrays", Journal of Spacecraft and Rockets, Vol. 24, No. 4 (1987), pp. 367-371.

Perez de la Cruz, C.; Hastings, D.E.; Ferguson, D.C.; Hillard, G.B. (1996). "Data analysis and model comparison for solar array module plasma interactions experiment," *Journal of Spacecraft and Rockets*, Vol. 33, Issue 3, pp 438-446.

Prag, A.B. (1983). "A comparison of the MSIS and Jacchia-70 models with measured atmospheric density data in the 120 to 200 km altitude range," NASA TR-0083 (3940-04)-1.

Purvis, C.K. (1985). "The Pix-II Experiment: An Overview," *Spacecraft Environmental Interactions Technology, 1983*, NASA CP-2359, AFGL-TR-85-0018, pp. 321-332, 1985.

Purvis, C.K.; Garrett, H.B.; Whittlesey, A.C.; Stevens, N.J. (1984). "Design Guidelines for Assessing and Controlling Spacecraft Charging Effects," NASA TP-2361, September 1984.

Reber, C. A.; Nicolet, M. (1965). "Investigation of the major constituents of the April–May 1963 heterosphere by the Explorer XVII satellite," Planetary and Space Science, 13, 617.

Reddell, B.; Alred, J.; Kramer, L.; Mikatarian, R.; Minow, J.; Koontz, S. (2006). "Analysis of ISS plasma interaction," AIAA-2006-0865, 44th AIAA Aerospace Sciences Meeting and Exhibit, Reno, NV, 9 - 12 January, 2006.

Reed, B.J.; Harden, D.E.; Ferguson, D.C.; Snyder, D.B. (2001). "Boeing's High Voltage Solar Tile Test Results." 17th Space Photovoltaic Research and Technology Conference, Ohio Aerospace Institute, September 11-13, 2001.

Rubin, A.G.; Stevens, N.J. (1983). "High voltage solar array models and Shuttle tile charging." AFGL Workshop on Nat. Charging of Large Space Struct. in Near Earth Polar Orbit, pp. 333-336.

Rubin, B.; Farnell, C.; Williams, J.; Vaughn, J.; Schneider, T.; Ferguson, D. (2009). "Magnetic filter type plasma source for ground-based simulation of low earth orbit environment." *Plasma Sources Science and Technology*, vol.18, 025015, 2009.

Rutledge, S.K.; Dever, J.A.; Banks, B.A.; Olle, R.M. "The Impact of Negative Grounding of Solar Arrays on the Sputtering of Array Surfaces in LEO", 1992 ASME/JSES/KSES International Solar Energy Conference, Maui, Hawaii, USA, April 5-9, 1992, published as *Solar Engineering 1992*, Vol. 2, pp. 811-816.

Samir, U.; Stone, N.H.; Wright, K.H. (1986). "On plasma disturbances caused by the motion of the Space Shuttle and small satellites – A comparison of in situ observations," *Journal of Geophysical Research*, Vol. 91, pp. 277-285, 1986.

Schneider, Todd; Hansen, Harold; Carruth, Jr., M. Ralph (2002a). "Minimum Arc Threshold Voltage Experiments on Extravehicular Mobility Unit Samples." AIAA 2002-1040, 40th AIAA Aerospace Sciences Meeting and Exhibit, Reno, Nevada, USA, 13-17 January 2002.

Schneider, T.A.; Carruth, Jr., M.R; Finckenor, M.M.; Vaughn, J.A.; Heard, J.; Ferguson, D. (2002b). "An Experimental Investigation of the Effects of Charging on the International Space Station." 7th Spacecraft Charging Technology Conference, Noordwijk, Netherlands, April 23-27, 2001.

Singh, N.; Samir, U.; Wright, Jr., K. H.; Stone, N. H. (1987). "A possible explanation of the electron temperature enhancement in the wake of a satellite," *Journal of Geophysical Research*, 92, p. 6100, 1987.

Singh, N.; Wright, Jr., K. H.; Stone, N. H., Eds. (1990). Current Collection from Space Plasmas, NASA CP-3089, December 1990.

Snyder, D.B. (1984). "Characteristics of arc currents on a negatively biased solar cell array in a plasma," NASA-TM-83728, July 1984.

Snyder, D.B.; Ferguson, D.C.; Vayner, B.V.; Galofaro, J.T. (2000). "New Spacecraft-Charging Solar Array Failure Mechanism." 6th Spacecraft Charging Technology Conference, AFRL-VS-TR-20001578, pp. 297-301.

Snyder, D.B.; Tyree, E. (1984). "The effect of plasma on solar cell array arc characteristics," NASA-TM-86887, January 1984.

Soldi, James D., Hastings, Daniel E., Hardy, David, Guidice, Donald, and Ray, Kevin, (1997). "Flight Data Analysis for the Photovoltaic Array Space Power Plus Diagnostics Experiment," *Journal of Spacecraft and Rockets*, Vol. 34, No. 1 (1997), pp. 92-103.

Stevens, N. J., Berkopec, F. D., Purvis, C. K., Grier, N., Staskus, J. V., (1978). "Investigation of high voltage spacecraft system interactions with plasma environments." AIAA 78-672, AIAA 13th International Electric Propulsion Conference, San Diego, CA, 25-27 April 1978.

Stillwell, R.P., R.S. Robison, and H.R. Kaufman, (1985). "Current collection from the space plasma through defects in solar array insulation," *Journal of Spacecraft and Rockets* 22 (6), 631-641 (1985).

Stone N. H., (1981). "The aerodynamics of bodies in a rarefied ionized gas with applications to spacecraft environmental dynamics," NASA TP 1933, November 1981.

Stone N. H.; Samir, U.; Wright, Jr., K. H. (1978). "Plasma disturbances created by probes in the ionosphere and their potential impact on Spacelab," *Journal of Geophysical Research*, 83, pp. 1668-1672, 1978.

Stone, N. H., Raitt, W. J., and Wright, Jr., K. H., (1999). "The TSS-1R electrodynamic tether experiment: Scientific and technological results," *Advances in Space Research*, 24, No. 8, 1037.

Szalai, K.J.; Bonifazi, C.; Joyce, P.M.; Schwinghamer, R.J.; White, R.D.; Bowersox, K.; Schneider, W.C.; Stadler, J.H.; Whittle, D.W. (1996). "TSS-1R Mission Failure Investigation Board," NASA-TM-112426, May 1996.

Toyoda, Kazuhiro; Cho, Mengu; Kawakita, Shirou; Takahashi, Masato (2008). "Investigation of Sustained Arc Under Solar Cell". 26th International Symposium on Space Technology and Science, Hamamatsu City, Japan, June 1-8, 2008.

Tribble, A. C., (1995). *The space environment – implications for spacecraft design*, Princeton University Press, ISBN 978-0691102993, 1995.

Upschulte, B.L.; Marinelli, W.J.; Carleton, K.L.; Weyl, G.; Aifer, E.; Hastings, D.E. (1994). "Arcing of negatively biased solar cells in a plasma environment," *Journal of Spacecraft and Rockets*, Vol. 31, Issue 3, pp. 493-501.

"U.S. standard atmosphere." (1976) NOAA. NOAA-S/T-76-1562.

Vaughn, J.A., (2003). "Plasma Interactions with a Negative Biased Electrodynamic Tether," 8th Spacecraft Charging Technology Conference, Huntsville, AL, October, 2003.

Vaughn, J.A.; McCollum, M.B.; Kamenetzky, R.R. (1997). "TSS-1R Failure Mode Evaluation." 31st Aerospace Mechanisms Symposium, Huntsville, AL, NASA CP-3350, pp. 309-320, May 1997.

Vayner, B.V.; Doreswamy, C.V.; Ferguson, D.C.; Galofaro, J.T.; Snyder, D.B. (1998). "Arcing on Aluminum Anodized Plates Immersed in Low Density Plasmas," *Journal of Spacecraft and Rockets*, 35, 6, pp. 805-811.

Vayner, B.; Galofaro, J.; Ferguson, D.; deGroot, W.; Thomson, C.; Dennison, J.R.; Davies, R. (1999). "The Conductor-Dielectric Junctions In a Low Density Plasma," NASA/TM-1999-209408, November 1999.

Vayner, B.; Galofaro, J.; Ferguson, D. (2001). "Arc Inception Mechanism on a Solar Array Immersed in a Low-Density Plasma," NASA/TM-2001-211070, July 2001.

Vayner, B.; Galofaro, J.; Ferguson, D.; Degroot, W. (2002). "Electrostatic Discharge Inception on a High-Voltage Solar Array," AIAA–2002–0631, AIAA 40th Aerospace Sciences Meeting and Exhibit, Reno, NV, January 2002.

Vayner, B., Galofaro, J., and Ferguson, D., (2003). "Ground Tests of High-Voltage Solar Arrays Immersed In a Low Density Plasma." AIAA-2003-4295, 34th AIAA Plasmadynamics and Lasers Conference, Orlando, FL, 23-26 June 2003.

Vayner, B.V.; Galofaro, J.T.; and Ferguson, D.C., (2004). "Interactions of High-Voltage Solar Arrays with Their Plasma Environment: Physical Processes," *Journal of Spacecraft and Rockets*, vol.41 no.6 (1031-1041), 2004.

Vayner, Boris; Ferguson, Dale C.; Galofaro, Joel T. (2008) "Emission Spectra of Arc Plasmas," IEEE Transactions on Plasma Science, Vol. 36:5, pp. 2219 – 2227.

Vayner, Boris V., (2015). "On Possible Arc Inception on Low Voltage Solar Array." AIAA-2015-4510. AIAA SPACE 2015 Conference and Exposition, Pasadena, CA, 31 Aug-2 Sep 2015.

Wright K. H. Jr., (1988). "A study of single and binary ion plasma expansion into laboratory-generated plasma wakes," NASA CR 4125, Feb. 1988.

Wright, K. H., Jr.; Swenson, C. M.; Thompson, D. C.; Barjatya, A.; Koontz, S.L.; Schneider, T.A.; Vaughn, J.A.; Minow, J.I.; Craven, P.D.; Coffey, V.N.; Parker, L. N.; Bui, T. (2008). "Charging of the International Space Station as observed by the Floating Potential Measurement Unit: Initial results," *IEEE Transactions on Plasma Science*, 36, Issue 5 Part 2, 2280, 2008.

www.ingramcontent.com/pod-product-compliance
Lightning Source LLC
Chambersburg PA
CBHW051159220526
45473CB00003B/830